P9-EGL-851

Too Small to Fail

How the Financial Industry Crisis Changed the World's Perceptions

Louis Hernandez, Jr.

AuthorHouse™
1663 Liberty Drive
Bloomington, IN 47403
www.authorhouse.com
Phone: 1-800-839-8640

First published by AuthorHouse 3/26/2010

ISBN: 978-1-4490-9068-5 (hc)
ISBN: 978-1-4490-9069-2 (e)

Library of Congress Control Number: 2010903405

Printed in the United States of America
Bloomington, Indiana

This book is printed on acid-free paper.

Praise for *Too Small to Fail*

"Louis convincingly builds the case that community financial institutions have a remarkable opportunity to redefine what has become a commoditized environment; retail banking. The global and large regional financial institutions are saddled with expensive and inflexible technology platforms. The community-focused banks and credit unions can redefine and reinvent the retail banking experience. He lays out a road map for such a transformation."
Butch Leonardson, SVP and CIO, BECU

"A unique analysis on the recent financial crisis from the technological perspective by an authoritative writer."
Jack Min Intanate, Founder and Chairman, DataOne

"Louis Hernandez is a rare visionary who understands the global implications of organizations that invest in innovation and those that don't."
William Mills, CEO, William Mills Agency

"Louis makes the case that successful banking institutions of tomorrow will need better technology to track, service and exploit the increasingly complex financial lives of the world's citizens. And further, he argues that the pace of social, economic and technological change is now so rapid that every day spent getting by on old technology rather than adopting fundamental innovation, puts a financial institution at a greater competitive disadvantage. Deep down, everyone in this business knows this to be true. Louis is sending the message that it's time to get serious about fundamental banking innovation."
Steve Post, CEO, Vermont Employees State Credit Union

"Overall, I found the book fascinating! It was a well-researched walk through time (and around the globe) that helped me better understand our current situation with actionable steps for what we should be doing today. I'm making sure everyone on our management team reads a copy."
Larry F. Tobin, President and CEO, *FAIRWINDS* Credit Union

"Louis Hernandez offers a compelling look at the financial services industry and the impact the recent financial crisis has had on the world. This innovator and business leader suggests opportunities that we should take in order to thrive in this new era. This is a must read!"
Christina C. Brown, President and CEO, Gesa Credit Union

"Louis provides the reader with a clear and concise explanation of the fundamental changes that occurred in the banking industry together with an equally understandable explanation of the pivotal role technology will play in the future evolution of the industry."
Brendan McDonagh, CEO, HSBC North America Holdings Inc.

To my family

Acknowledgement

Writing a book involves countless hours of research, writing and editing. I offer my sincere gratitude to a number of individuals whose commitment helped make this book possible, including: Kimberly Gerdis, who managed the research effort and other aspects of the book's publication; Pat Bator, who assisted with the research; Bill Handel, who provided invaluable industry expertise; Charlie Beer, Ted Fajardo, John Messier, David Mitchell, Lizette Nigro, Sue Pinsonneault and Ron Young, who provided input based on their extensive expertise in the financial services industry; and my assistant Robyn Tjornhom, without whose dedication this book would not have been possible.

I'd also like to thank several industry experts and business leaders for their input, including: Christine Barry, research director, Aite Group; JuliAnne Callis, president, National Institutes of Health Federal Credit Union; Dennis Dollar, principal partner, Dennis Dollar Associates; Butch Leonardson, chief information officer and senior vice president, Information Technology, BECU; William Mills, CEO, William Mills Agency; and Steve Post, CEO, Vermont State Employees Credit Union.

To all of you - and to others who offered input - thank you for sharing your time and expertise.

Last, but certainly not least, I want to acknowledge the many sources cited in the book who so kindly allowed me to use their research.

Thank you.

Contents

Preface

What a dramatic set of changes we have all been through as the global economic engine adjusted to new realities. The credit crisis triggered by the subprime mortgage market in the United States quickly spread to other credit products, and it ultimately threatened not only the U.S. banking industry but also the global economy. This near-disaster impacted financial institutions throughout the world and changed the banking landscape in an unprecedented fashion through consolidation, bailouts and new government regulations, to name a few.

Even before the financial crisis exploded onto the international scene, seismic shifts in the geopolitical, economic, demographic and technological arenas were impacting the global topography and, as a result, the broader financial services market. While these changes represent a continued evolution of how we live, work and play, their acceleration has dramatically impacted the role of financial institutions.

Meanwhile, the dramatic shift in the financial services landscape, which reached a critical point in 2008, masked some larger issues faced by the industry. As the competitive landscape intensified, boards and executives were increasingly looking for new ways to compete and new strategies to offset fundamental shifts in the basic business model for financial institutions.

The increasingly global competitive environment for financial institutions has resulted in accelerating margin compression, channel proliferation, pressure to increase non-interest fee income, increased regulatory scrutiny, heightened fraud and security risks and the "consumerization" of the entire industry. To stay competitive, financial institutions must diversify revenue streams, increase non-interest fee income, know more about their account holders, capitalize on the increased interaction with the people they serve through more touch points, address fraud, security and regulatory concerns with greater flexibility, all while dramatically improving operating efficiency.

Rather than addressing these issues directly, many in the industry, through a combination of factors, chose to pursue less strategic, but more profitable, product and service strategies. The accelerated pace of earnings that resulted only reached a critical point when the subprime market began to falter. Because of the subsequent shakeout in the financial markets, the strategy of pursuing high-return/high-risk profile products as a way to mask the underlying business model shift has decreased. Now that this has been exposed, the pressure will once again return to the more fundamental issues that need to be addressed by the industry; particularly, how it will survive and thrive for the future. As signs of stability emerge, institutions will have to provide greater clarity about their contribution to the economic engine and redefine a long-term sustainable business model.

At the same time, financial institutions, which once pioneered computational innovation, are now stuck with some of the oldest technologies of any industry, layering on and masking this weakness with newer technology in a desperate attempt to meet their strategic needs. This costly and outdated technology infrastructure has been a nagging issue, exacerbated by inertia and regulatory incentives against change. It has turned into a severe competitive disadvantage as institutions look for greater flexibility and efficiency in meeting the changing landscape of the industry. By masking these legacy systems with layers of newer technologies, the industry's infrastructure has become expensive, inflexible and increasingly ineffective.

This became evident as industry pressures mounted, and poorly integrated, outdated, and inaccurate information contributed to the disjointed nature of the subprime and related credit and risk decisions. Large institutions thought to be more stable with their scale benefits, diversity and capital bases proved to be, in large part, the weak link as the economic environment became more difficult.

The "Too Big to Fail" thesis may give way to the less diversified, less scaled but seemingly more prudent community-based institutions that largely avoided the subprime crisis by maintaining direct ties to their communities. This collaborative network of community-based institutions shows renewed value, reflecting a relatively unique pillar of stability for our economy and a new mantra of "Too Small to Fail."

This book reflects on the pace of change that impacts all aspects of how we live, its implications on the financial services industry, and the unique interrelationship of recent history and the current state of the financial industry. It also lays out a path to navigate the future that will ultimately reposition the industry and individual institutions for long-term success.

Note from the Author

I have been in the financial industry in one way or another my entire career, but came specifically to the banking segment just over a decade ago when I was asked to lead a small startup company, Open Solutions Inc. (We were just over $10 million in revenues at the time and we lost more than that my first year here.) Having worked with a few high growth companies before, I knew success was usually dependent on a few key people who would make all the difference. Open Solutions was no exception. It took a special effort from our founders, early investors, initial clients, and employees to get things going. Our regulators did their part as well, and helped educate me on the reasons why there were very few enterprise software startups in the history of our regulated market space.

As I worked with one of our key founders, Cliff Waggoner, who was a relational and data modeling expert, I began to appreciate the profound impact his approach to managing relationships through technology could be on our industry. His innovative perspective would have a dramatic impact on making the human interaction in banking more powerful and at a very low cost.

I soon hit the road to meet and learn more about the needs of the industry and our clients. Our very first client was Simsbury Bank (still with us today), a commercial bank located in Simsbury, Connecticut. Barry Loucks was starting up the de novo as its first president. Like a lot of the bank and credit union leaders I've met over time, he had deep roots in his community. He grew up in the area; went to the University of Connecticut; raised his family there; developed his professional resume by working at the well-known, large local banks; and was named man of the year by various local organizations more than a few times. He also understood the role a bank played in the community and how technology could be used as an enabler of strategy. He had a lot of experience throughout his banking career with many of the legacy technology tools that are still in place today, and was always willing to look for better ways to run a bank. Starting a new bank using a brand new enterprise system and technology that no other provider had tried before was a very big chance for this bank president to take, particularly as he was raising money, getting regulatory approval and hiring employees all at the same time.

Working tirelessly with these creative technologists and pioneering bankers, we created a shared vision for a better tomorrow that began in the great American tradition of a pure startup. Soon we became recognized as one of the fastest growing software companies in the world, having partnered with some of the most innovative financial institutions of all sizes, charters and locations, first in the U.S., then North America and finally, around the world. We became a publicly traded company, achieving exceptional success considering our unique and humble beginnings.

Even with these many accomplishments, we wanted to do more. So we took the company private, at a valuation of almost $1.5 billion, to better prepare ourselves to meet the requirements of the quickly changing financial industry for the next 15 years. With the assistance of our partners, we spent almost $80 million on our transformation to create a more directed service organization with a refreshed infrastructure and product suite to help everyone in the world take advantage of what we were doing.

So why did I feel compelled to write this book? While the industry has treated us well, more inspiring to me was the faith and trust these institutions and their leaders continue to place with us, particularly since this is such a conservative industry. This always had a profound impact on me as we literally had zero clients at one point. We relied heavily on this partnership and shared vision approach to address the industry's needs together. Over the years, I've met with thousands of financial institution leaders. One by one. In board rooms and at conferences. In airports and hotel lobbies. In dining rooms and downtown offices. All around the world. They believed in our vision, and they're still standing with us as we move ahead. I still believe in their passion to truly "serve and support" their local communities and markets. In the process, they have become my extended family.

The financial industry is going through a number of essential changes impacting its business model and competitive positioning. Some of the basic tenants of the need for a trusted, appropriately regulated financial intermediary are as strong as ever. The financial sector may be the ideal industry to not only stabilize the markets, but also provide the prescription for a healthy global economic engine. This need for a trusted financial intermediary focused on local community needs in an increasingly complex and global economic landscape creates a fantastic opportunity to not only address the business model dynamics in the industry but also fulfill an important social and economic role that was part of the industry's foundation. This will force institutions to be clearer about how they differentiate themselves and make that difference more obvious with every interaction. The future path will likely embrace the values of the past. This "back-to-basics" approach will have to reflect the new realities and will require extraordinary leadership.

Courageous, thoughtful leaders among us will have to emerge to show the way, commit to their communities and their institutions, and embrace the core values of the industry while addressing a very different operating backdrop. This call to arms for the best and brightest will be more important than the resolution of the recent financial services led economic trends in the sense that it will set the stage for the future of the industry.

Today, I find an industry that is at a crossroads. I feel compelled to reinvest in our industry and join with other leaders to find a way to a better tomorrow. Through my service of just about 4,000 financial institutions and countless others that I consult with, I feel not only that we can navigate through this period together, but emerge, recapturing the energy and vitality that made many of us so proud to be part of the financial services family.

Louis Hernandez, Jr.

Chairman and CEO
Open Solutions Inc.

1: Accelerating Global Change

"Things do change. The only question is that since things are deteriorating so quickly, will society and man's habits change quickly enough?" – Isaac Asimov

---

The last few years have been fraught with tremendous changes in the global economy, particularly in the financial services area. I've attended planning sessions for financial institutions of all sizes in many parts of the world over the last decade, and I'm struck by how frequently discussions on the rapidly changing market conditions for our industry are highlighted with exasperation. I'm also struck by how our industry so often mirrors the changing human dynamic. In the last two years, I've noticed uncertainty and a significant shift in attitude as very few in the industry today have lived through a crisis that was this broad and severe.

The initial sense of fear and foreboding has eased as relative stability emerged against an otherwise stark backdrop. Leaders and institutions are more willing to think carefully about what these changes mean. They may finally address long-standing industry issues that could have been avoided if it weren't for the operating environment allowing enough flexibility to pass these issues onto the next generation as leaders battled today's issues.

Who would have guessed that some of the largest and most trusted financial institutions in the world would have disappeared in such dramatic fashion? Or, that incumbent President George W. Bush would go from a solid re-election margin in 2004 to approval ratings in the 20 percent range before the next election cycle got underway? Or, that the U.S. would elect its first-ever black president? Or, that Asia would be perceived to be the strongest and most reliable economic growth engine in the world? In addition, who would have imagined that North Korea and Iran would provide the most significant near term global nuclear threats? Or, that Google and smart phones would have such an impact on how we live?

And if it seems like things continue to change at a faster rate than ever before, it's not because you are out of touch. Research shows that the pace of change is accelerating in many dimensions.

Let's use technology as an example. Technology has always enjoyed a prominent place in the annals of history because of the significance of innovation. Examples of this progress include electricity, computers, and more recently, personal communication devices. The rate of change for technology acts as an accelerator for constructing the way we live because it is so intertwined with who we are. As a result, technology has an important role in shaping the geopolitical and economic landscape, demographics, commerce and even religious beliefs to create an increasingly fast-paced and complex world.

Basic technological advances such as fire, the wheel, and creating and leveraging tools evolved over thousands of years. And they had a profound impact on how humans lived. Technological breakthroughs began accelerating in the 19th century to the point where we saw more technological change than in the nine centuries preceding it. Then, in the first twenty years of the 20th century, we saw more advancement than in all of the 19th century. Researchers refer to the "paradigm shift rate" (i.e., the overall rate of technical progress) as picking up speed at an incredible rate – doubling in speed every decade or so. The technological progress now taking place in the 21st century would have taken 20,000 years in the past. Even when compared to the 20th century, the 21st century will see a thousand times greater technological change than the previous 100 years. Of course, Gordon Moore, co-founder of Intel, made this principle even more popular when in predicted in 1965 the doubling of computing power every two years.[1] Many prognosticators have shortened this window to 18 months and predict that this trend will continue to hold for the next 20 years or so. By that time, transistors will be only a few atoms thick.

Now, paradigm shifts occur in only a few years time. From electricity to the assembly line, to the telephone and now the Internet, digital music players, electronic books and Personal Digital Assistants (PDAs) offer the latest examples of accelerated changes. And this has a direct impact on the commercial environment. For example, e-commerce represented 0.6 percent of total retail sales just 10 years ago. That's approximately $4.6 billion compared to 3.6 percent, or approximately $32.4 billion today, an increase of more than 600 percent.[2]

In addition, these advances entwine themselves with the economic, social, demographic, and political dynamics. Take one of the greatest retail success stories of the last 50 years: Walmart. An essential ingredient in Walmart's success is its swift embrace of the latest technology. Almost from the beginning, Sam Walton recognized that the solution to keeping costs down and profits up was rigid inventory control. This meant ordering just the right items in just the right amounts, thereby improving his inventory turnover ratio. As a result, Walmart standardized a now common industry practice of minimum re-order points and consolidated supplier information to maximize negotiating positions with vendors. In the process, it has created an inventory control and ordering system that is the envy

1 Ray Kurzweil, *The Law of Accelerating Returns* (2001), http://www.kurzweilai.net/articles/art0134.html?printable=1.

2 U.S. Census Bureau, *Table 3. Estimated Quarterly U.S. Retail Sales (Adjusted1): Total and e-commerce2*, U.S. Census Bureau, http://www.census.gov/retail/mrts/www/data/html/09Q2table3.html.

of the retailing industry and an information repository that ranks among the largest commercially available.[3]

Not only are technological and economic changes taking place at a phenomenal rate, but social transformations are happening at a more rapid pace as well. I've been impressed by how much of an impact financial institutions have in communities around the world. From small towns in the mid-western part of the U.S., to villages in Thailand or Vietnam, to large cities like New York, London and Hong Kong, financial institutions continue to represent the very fabric of how we live and interact. The types of loans issued, deposit trends, delinquency rates, methods of interaction, and even regulations highlight the trends of people and communities, their struggles and their triumphs.

But before we turn to banking, let's review a couple of macro-trends as we reflect upon the changes in our own industry and their impact on our communities. What emerges are some far reaching economic, geopolitical, demographic and workforce changes that impact each of us no matter where we are geographically. By understanding these major themes, we'll be in a better position to understand their implications to the financial industry, which will serve to create the basis for assessing the risks and opportunities that lie ahead.

Economic Trends

The initial signs of financial market challenges have given way to more significant changes in the global economic climate, resulting in what many refer to as the Great Recession. Financial institutions need to understand what long-term implications this may have on the people they serve, to decide how to navigate forward.

---

3 "Sam Walton: Bargain Basement Billionaire," *Entrepreneur*, http://www.entrepreneur.com/growyourbusiness/radicalsandvisionaries/article197560.html.

The current economic situation often raises comparisons to the Great Depression. The circumstances that led to this major economic downturn sound familiar. Most economists continue to debate the cause of the Great Depression. Some suggest a spending hypothesis where there was a fall in spending on goods and services. Some argue that the stock market crash of 1929 caused a shift by reducing wealth and increasing uncertainty about the U.S. economy's future. Others believe the down turn was a result of a drop in real estate investment triggered by overbuilding in the 1920s.[4] Milton Friedman and Anna Schwartz believe that contractions in the money supply are responsible for most economic downturns and the Great Depression is "a vivid example."[5]

There are clearly similarities between the Great Depression and today, but there are differences as well. The current economic climate bears disturbing similarities to the start of the Great Depression, such as the huge drops in the stock market that reduced people's wealth and decreased spending, and a banking system crippled by bad loans and speculative real estate.

Many economists feel intervention by the world's central banks, including The Federal Reserve, may be the biggest difference in thwarting a repeat of the Great Depression. Time will tell if the government and regulatory actions helped or exacerbated the situation.

Another commonly highlighted difference between the Great Recession and the Great Depression is the shift from an agricultural and manufacturing based economy – as it was during the Great Depression – to an economy that is more reliant on services and consumer spending, as it is today. In addition, the globalization of

4 N. Gregory Mankiw, *Macroeconomics*, (New York: Worth, 2003), 350-365.

5 Milton Friedman, *Capitalism and Freedom*, (Chicago: University of Chicago Press, 1962), 7-17.

the economy and money supply are much different than they were in the Great Depression.

We'll also see whether this financial crisis has the same kind of lasting psychological impact that the Great Depression had on human behavior. Either way, the current relative stability is a welcome change for most people.

With the U.S. savings rate reaching an annualized 6.9 percent – the highest in 15 years – the nation is seeing a huge cultural shift in consumer behavior. As consumer confidence and incomes begin to increase, we should see a boost in consumer spending.

At the end of 2009, Americans showed growth in spending while also continuing to save at the highest level in six months, according to the Commerce Department's Bureau of Economic Analysis. Both are good signs and we'll see if they are sustainable. Some economists suggest that having bigger cash reserves frees U.S. reliance on foreign investment to generate economic expansion. As Bloomberg reported, "The current-account deficit, which includes trade in goods, services and income transfers, narrowed in the first quarter [2009] to its lowest since 2001 as Americans saved more and brought fewer imports."[6]

"[However], incoming data also point to America's sustained and perplexing dependence on foreign capital inflows - a dependence that suggests an underlying economic vulnerability that has yet to be addressed. Whether it needs to be addressed next month, next year, or next decade is still a question that continues to haunt the followers of global macro trends," offers Tim Duy in his column FedWatch.[7]

---

6 Rich Miller and Alison Sider, "Surging U.S. Savings Rate Reduces Dependence on China," *Bloomberg*, June 26, 2009 http://www.bloomberg.com/apps/news?pid=20601109&sid=aomel_t5Z5y8.

7 Tim Duy, comment on "A Tangled Policy Web," Tim Duy's Fed Watch, comment posted on June 2009, http://economistsview.typepad.com.

The nature of our industry often forces us to become students of economic data. What will become increasingly important to understand is the impact of the current economic trends on the psyche of the people we serve, their attitudes toward risk and spending, and the social implications on communities. The financial models, pricing and packaging, and assessments of risk and return will be impacted by these implications.

Highlighted below are some key economic U.S. data that directly impacted our industry at the depth of the Great Recession:

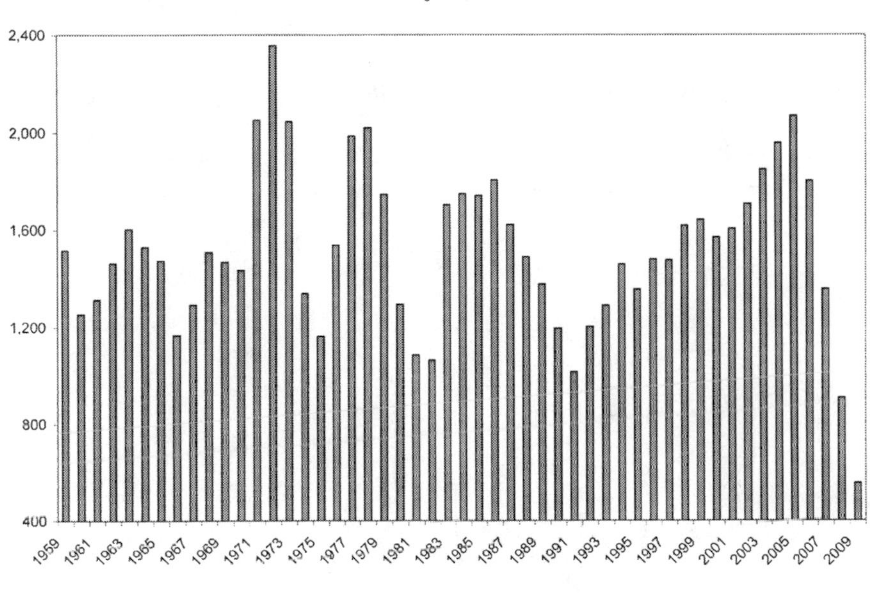

Housing starts were at historic lows in 2009, the fourth year of a steeply declining trend. This decline followed a period of sustained growth in housing starts dating back to the early 1990s.

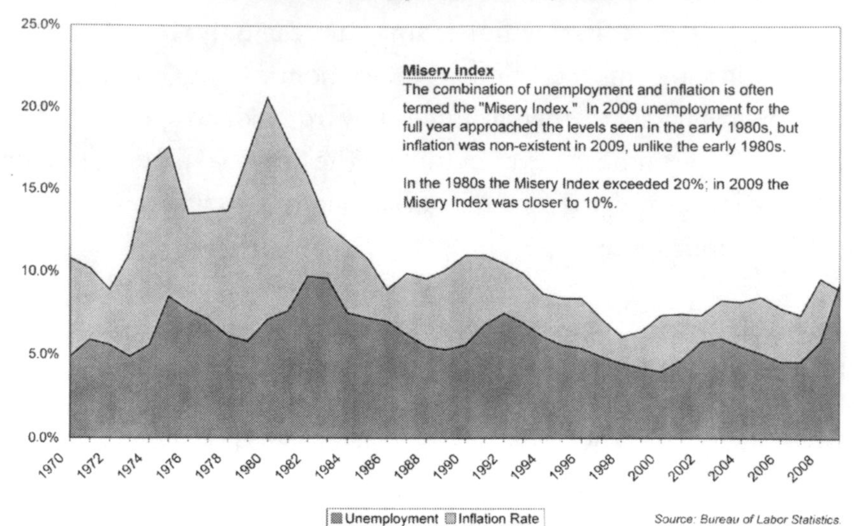

The Misery Index is often used to measure the economic distress being felt by consumers. It is measured as the combination of the unemployment rate and inflation. Our level of "misery" in 2009 reached a high point for the decade, but falls substantially below the misery felt in the 1980s.

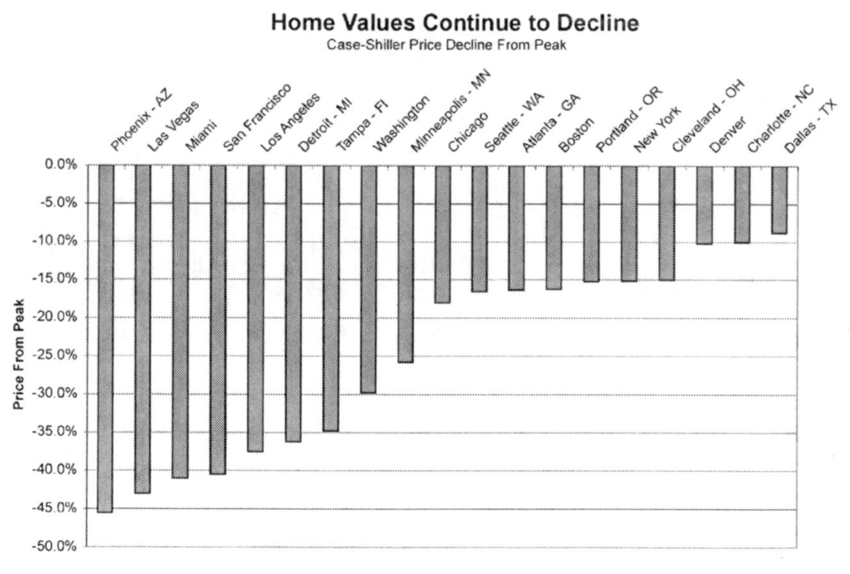

A major factor behind the recent economic debacle is declining home values. Declines have been felt in virtually all markets, but the chart illustrates that some areas were significantly more impacted than others. In particular, markets in the "sand states" (California, Florida, Nevada and Arizona) experienced the highest declines in home values.

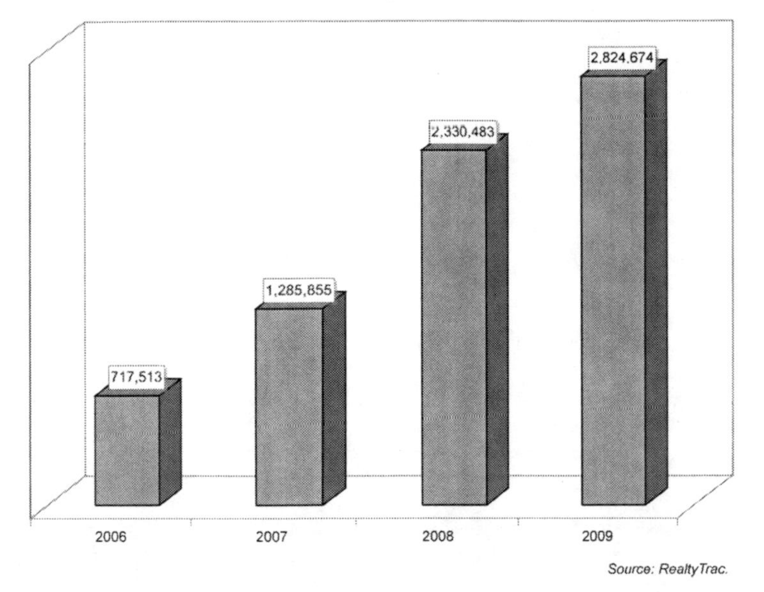

Foreclosures have been a significant contributor to this economic crisis and to industry problems. Foreclosures continued to climb in 2009, although at a slower rate.

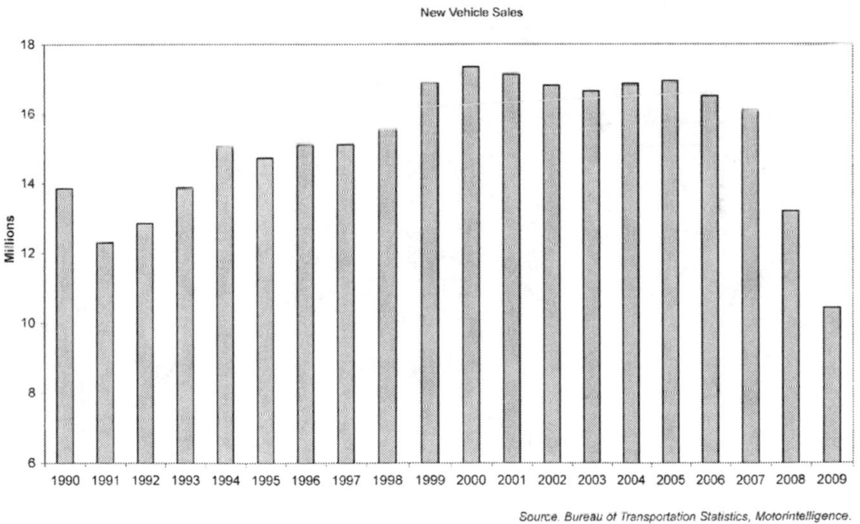

Automobile sales are a key component of economic performance. Automobile sales were at historically low levels in 2009. There is significant pent-up demand that will release when the consumer begins to feel as though the economy is in recovery.

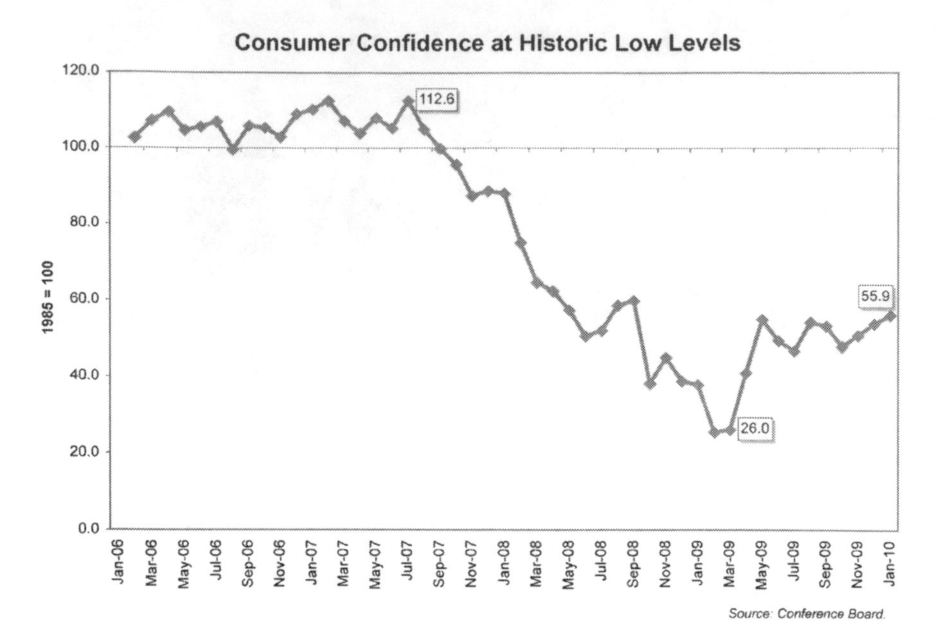

Consumer confidence also reached historic lows in 2009. There has been improvement in this index, but it still remains significantly below historic norms. Consumer confidence is critical because it drives consumer spending.

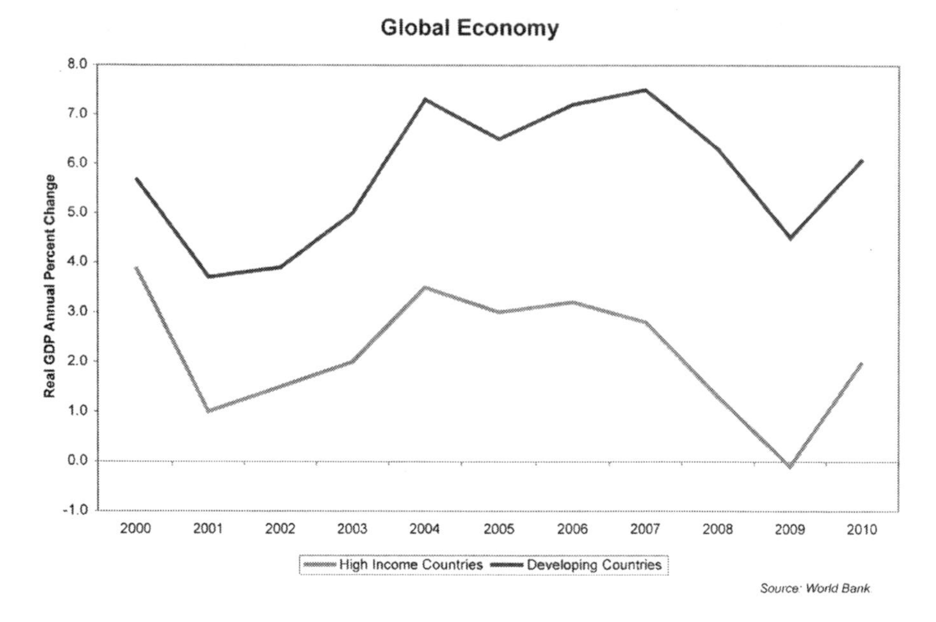

The chart illustrates the interconnectedness of the world's economies. The level of global trading and capital exchange makes this truly a world economy.

The long-term impact of these dramatic changes is yet to be seen.

Accelerating Globalization

The steady, increasing globalization that has been accelerating for decades became even more significant as many former communist countries began to adopt capitalistic ideals in various forms. This opening of new markets, the increased rate of our own interactions helped by technology, and the drive for global expansion led to the global village we all participate in. All economies must now grapple with this heightened level of interconnectedness. It is shaping the future global economic engine and has forced everyone to rethink its global role and sphere of influence. Capitalism in its various forms continues to evolve around the world and has had an impact on the economic changes.

The steady march toward capitalism started when prehistoric civilizations realized that farming and ranching could be profitable.[8] The connection between cattle and money explains the common root of the words cattle, chattel and capital. It was during the Greek and Roman dynasties that trading and market economies truly started to thrive. Often goods were exchanged in trade because coinage varied in value, consistency and availability.

It seems that the early forms of capitalism and what some refer to as the great capitalism "experiment" discussed by the Greeks and Romans and debated by their predecessors, exists in some form through much of the world today.

The interconnection between capitalism and freedom is more apparent as new worldwide economic paradigms have emerged that will characterize international relations over the next few generations.

---

8 Donald L. Hamilton, *The Capitalists*, (1999 -2006), http://novan.com/capitlst.htm.

The capitalist dance forced by this sensitive relationship is shaping various forms of capitalism around the world. Capitalist-authoritarian societies with their booming economies are taking on their liberal-democratic counterparts. China and Russia are at the front of this alleged authoritarian bloc. During the Cold War, the competition was between capitalism and communism. China and Russia have adopted capitalism in one form or another.

The fall of the Berlin Wall, transformation in Russia, and China's expansion of its economic model to be more open gave birth to models combining human rights, civil liberties, freedom and capitalism in different parts of the world, including former communist countries. Their early success in maintaining a strong authoritarian model blended with some capitalist elements may slow the adoption of Western-based principles. Or it may lead to a slow, steady adoption of common views for a global marketplace and a standard of freedom throughout the world. While there are varying economic models, we seem to now be linked in ways that never previously existed.

A recent study from the International Monetary Fund showed that growth in so-called 'capitalist' authoritarian countries was three times higher than liberal democracies.[9] With this rate of growth, authoritarian countries may be feeling good about their performance, and therefore slow to adopt more liberal ideas. What is not clear is if this growth is a function of their authoritarian model, or due to the industrialization of former underdeveloped countries.

It is important to note that this is occurring with less talked about, formerly underdeveloped countries as well. Poland, for example, is the European Union's largest accession nation with growth potential.

---

9 John Lee, "Western vs. Authoritarian Capitalism," *The Diplomat*, June 18, 2009, http://www.the-diplomat.com/001fl281_r.aspx?artid=134.

Poland's real GDP increased by 54 percent and its manufacturing output more than doubled from 1991 to 2000.[10]

Since then, the Polish economy has not slowed down. A 2008 Forrester report indicates, "The income of the average Polish consumer is growing and will increase even more in the future, thanks to a developing economy and high levels of investment in the region."[11]

We have never had a situation with this degree of global interconnectivity. Many citizens of the world were surprised to see the significance of the impact the U.S.-led slowdown had on the rest of the world. Naturally, the large U.S. economic engine was known to have great impact, but many did not realize the degree to which the world was integrated to each other's fortunes. It is hard to predict how profound the impact of this economic slowdown will be on the rest of the world over the long-term. Either the interrelationship of nations will precipitate a domino effect that will negatively affect all economies, or the robust growth among developing countries will mitigate the developed-country deceleration. This may become one of the most important learnings from this downturn; and could shape how aggressive governments encourage continued globalization through public policy.

The banking industry must consider the changing economic landscape as the roles and their implication on global commerce are also impacted. While we have experienced significant economic change in the past, today's increased level of interconnectivity could impact even the smallest communities and should be monitored.

---

10 Boleslaw Domanski, "Industrial Change and Foreign Direct Investment in the Postsocialist Economy: The Case of Poland," *European Urban and Regional Studies 10* (2003). http://econlab.uom.gr/~move/images/stories/articles/domanski_industrial_change.pdf.

11 "Poland 2007: Emerging Market of Technology Optimists," *Forrester* (March 2008).

The Economy and the Hierarchy of Needs

The changing regional hierarchy of needs also has repercussions on the new global economic landscape. The reason this can be so important is because as a regional group's needs change, what they are willing to do to meet their needs and its impact on the increasingly interconnected economic landscape can be significant. It can impact labor rates (as individuals in regions decide what they are willing to do for what price), education levels, buying patterns and even productivity rates.

In his 1943 paper, *A Theory of Human Motivation* and his subsequent book *Motivation and Personality*, psychologist Abraham Maslow first introduced his concept of a hierarchy of needs in which he suggests that people are motivated to realize basic needs before moving on to other needs. That is, the need for food, water, sleep and warmth. Once these needs have been fulfilled, people move to the next subsequent stage of needs, such as safety and security.[12]

Michael R. Hagerty's cross-study published in *Social Indicators Research* attempted to understand how and if Maslow's hierarchy of needs affected countries throughout the world. In his study, Hagerty proved that individuals learn successfully over time, thereby fulfilling their lower-level needs and progressing on to higher-level needs. Additionally, the study concludes that once individuals are able to survive in a country, opportunities arise for them to thrive.[13]

This is important, because while countries need political and economic opportunity, they also need an intellectual curiosity to stay competitive and an ability to develop new ideas and products. In Mexico, India, and other undeveloped economies, there were

12 Abraham Harold Maslow, *Motivation and Personality* (Harper Collins Publishers, 1987).

13 Michael R. Hagerty, "Testing Maslow's Hierarchy of Needs: National Quality-of-Life Across Time," *Social Indicators Research* 46 (1999).

significant geographic areas and population centers without education, employment, infrastructure, or housing. As basic food and shelter was provided, the populace demanded more social services; and many of these necessities are now being provided in some form in concentrated locations.

Keep in mind that some of the emerging economies will have other issues that could inhibit them from becoming a more integrated part of the growth network with other countries. Cultural barriers and hierarchy of needs can influence where you end up in the global economy and whether your advances are temporary or long-term. A person's willingness to work for a certain value may diminish as their desire for basic needs, and willingness to expend time and effort to meet those needs, change. As their basic needs are met, they may begin to pursue higher order needs that require higher pay, particularly if their training levels are equivalent. For example, an IT or farm worker's willingness to do the same work at a lower rate may change, as their individual needs change. This has to factor into how countries or geographies manage their competitive profile and business outlook in areas such as labor costs and rates of employment.

Economics and Education Linked

Economics and education are major dynamics at the core of this global transformation.

In the year 2000, the U.S. accounted for 31 percent of world GDP, with China representing 3.7 percent. By 2009, U.S. share of world GDP was projected to have fallen to 24.9 percent, while China's share grew to 8.3 percent.[14] Further illuminating these trends, in 2000 per capita GDP in the U.S. was $39,300, and it grew to $41,700 in 2009.

14 International Monetary Fund, *World Economic and Financial Surveys: World Economic Outlook Database*, http://www.imf.org/external/pubs/ft/weo/2009/02/weodata/index.aspx.

In China, by contrast, GDP in 2000 was $1,100 per capita, but by 2009 had grown to $2,500. Obviously, China's GDP per capita pales in comparison to U.S. figures, but in that nine-year span it more than doubled, compared to six percent growth of GDP per capita in the U.S.[15]

Some aggressive projections indicate that China will overtake the U.S. in terms of economic output within a decade, according to estimates released by Deutsche Bank.[16] Expecting to fuel China's growth is the rapid expansion in emerging market economies, which will account for about 70 percent of global GDP growth in the coming decade.[17]

With regard to education, the United States no longer dominates. It is not so much that the U.S. has lost its edge – the fact remains that the U.S. percentage of college graduates continues to rise annually. But other countries have raised the bar and are becoming more competitive.

Norway, Britain, and the Netherlands climbed past the U.S. in a number of education evaluation categories several decades ago, according to a 2000 report by the Organization for Economic Cooperation and Development, which pooled data from its 29 democratic countries and 16 nonmember countries.

In fact, graduation statistics in Britain went from less than 20 percent in the early 1990s to 35 percent by 2000. In the Netherlands, 34.6 percent of the student population graduated from college, and in

---

15 United States Department of Agriculture: Economic Research Service, *International Macroeconomic Data Set*, http://www.ers.usda.gov/Data/Macroeconomics/.

16 Chris Oliver, "China's GDP to overtake U.S. by early 2020s, says analyst," *The Wall Street Journal*, April 23, 2009, http://www.marketwatch.com/story/chinas-gdp-overtake-us-early.

17 Visualizing Economics, "Share of GDP: China, India, Japan, Latin America, Western Europe, United States," January 20, 2008, http://www.visualizingeconomics.com/2008/01/20/share-of-world-gdp/.

Norway the figure rose to 37.1 percent. In addition, many countries surveyed showed enrollment improvements, some by more than 20 percent whereas the U.S. only had a 3 percent increase.[18]

While the U.S. has made some significant investments in hard sciences education and encouraged innovation, the investments pale in comparison to China, which has more than five million graduates a year in secondary schools alone. China is currently graduating 600,000 people with science and engineering degrees, primarily at the undergraduate level. Over the next decade, that number could reach more than 5 million.[19] Based on these projections, it is certainly possible that China could outpace the United States by 2020, despite the renewed American focus on research and development.

Another way to look at these numbers is that one quarter of the population in China with the highest IQ, is greater than the total population of North America. (In India, the top 28 percent exceeds the North American population totals).

Whatever the contribution education had to growth in the past, investments in human capital (education) will likely rise in importance relative to investments in other forms of capital as the transition to a post-industrial, knowledge-based economy accelerates.

In India, for example, output grew some 3.5 percent over the last 40 years, while an increase in labor productivity, a major indicator of growing wages and standard of living, climbed 2.4 percent per year.

---

18 David Abel, "Going Backwards: US Falls Behind In College Graduation Rate," *Boston Globe,* May 17, 2000.

19 Shahid Yusuf et al., "China and India Reshape Global Industrial Geography," in *Dancing with Giants: China, India and the Global Economy,* ed. L. Alan Winters and Shahid Yusuf (Singapore: The World Bank and Institute of Policy Studies, 2007), http://siteresources.worldbank.org/INTCHIINDGLOECO/Resources/CE_Ch02pp.027-56_FINAL.pdf.

It is believed that education contributed from between 13 percent to 30 percent of the total labor productivity increase.

The obvious connection between investment in education and a return in improving economic development is no surprise. Education ensures a skilled work force that needs to be the fulcrum of development for raising the quality of life in developing nations. Investment must be recurrent and meaningful.[20]

Whatever the status, change must take place in the U.S. educational system in order for the nation's youth to remain competitive in the future. This education equation will be central to how the economic landscape takes shape and how regional strengths emerge.

It is difficult for the world as a whole to avoid being impacted by these trends. India and China have very large populations and are undeterred in their industrialization. The result will be continued rapid gains in both nations, much like other industrialized nations have already enjoyed.

More important than any act of government or series of laws will be the will of the people. The test will be how they use their available freedoms to produce a better standard of living and a more innovative culture to take advantage of the changes today's world offers and tomorrow's world promises.

The Geopolitical Landscape

While the current focus is on the economic landscape, it is hard to have a complete discussion without assessing the implications of the geopolitical landscape. As we grapple with the world's changes, as

---

20 Rudra Prakash Pradhan, "Education and Economic Growth in India: Using Error Correction Modelling. International Research Journal of Finance and Economics," *International Research Journal of Finance and Economics* 25 (2009), http://www.eurojournals.com/irjfe_25_11.pdf.

well as those in our local communities, many wonder just how much change we can endure. We've discussed that the rate of change is clearly accelerating and that the economic changes, globalization, our interconnectedness and changing hierarchy of needs will manifest themselves in some way in the people that we serve. Equally important, are the geopolitical changes around the world and how they influence policy and shape behaviors in every community.

In times of geopolitical change, previously obscure groups often replace the old guard. They can bring a fresh vitality to lackluster and tired societies. The same holds true in companies and other affinity groups.

While there is ample basis for discussion about current global powers, it is hard to know where the West might be in its historical influence in world affairs. What seems quite clear is the West's role and influence is changing. Coping with the change may be more difficult than the reality of the situation.

How much of the current economic shifts are due to shifting dynamics of emerging nations, the renaissance of former communist countries or of the shifting educational population around the world won't be known for some time. But it may be setting the seeds for the next phase of global influence around the world.

Questions proliferate about the historical influence centering on the rise of China and India, the growth and influence of Brazil, and the reemergence of Russia. Although the primary power shift view is often from a militaristic perspective, it may be more appropriate to look at it from an economic or commercial perspective.

For instance, the early empires were centered around areas of commerce (i.e., the cradles of civilizations) such as the Tigris-Euphrates Valley, Indus River, Nile River, Yangtze River, and later the Amazon. They were the centers in the transportation of goods and natural resources (commodities) to and from the hubs of civilization – providing those commodities of wealth to the individuals, dynasties, or eventually governments held in power by strong militaries.

As human mobility and advances in transportation and communication accelerated, so did the methods of commerce. This has had a direct impact on the commercial centers' access to and use of goods and services. From water and caravans, to trains and trucks, to air and cyberspace, the shifts and ease of human interaction, information, and knowledge have changed dramatically. Therefore the tools for competition have also changed.

There appears to be a natural lifecycle of kingdoms, civilizations, and empires. History shows that they rise and fall, prosper and decline, succeed and fail. However, some last longer than others. The Roman, Greek, Siam and English Empires, the Great Chinese Dynasties, and the Aztec and Mayan Civilizations were once powerful realms. With their influence came the opportunity to control commerce and currency.

There are lessons in the rise and fall of two superpowers of their day: the Han Dynasty (206 B.C.–220 A.D.), which reigned for more than 400 years, and the Roman Empire (31 B.C. – 476 A.D.), which lasted about 500 years.

The Han Dynasty introduced new and momentous inventions including the invention of paper and gunpowder. It also opened up the Silk Road trade route that moved goods ranging from exotic animals to precious stones, while simultaneously serving as a vehicle for the spread of knowledge, ideas, culture and religion.

The Han Dynasty paved the way for another prosperous era, the Tang Dynasty (618 to 907 A.D.), during which the Silk Road continued its progress. According to The Silk Road, by Dr. Oliver Wild, the city of Changan was the starting point of the route and its population grew to almost two million — including five thousand foreigners — during the Tang rule.

This area became an integral part of trade between the East and West. More importantly, the development of the Silk Road spread Buddhism from China to India and enabled culture to flow without restraint between the two.

While the Han Dynasty flourished, Augustus Caesar also transformed the Roman Republic into the Roman Empire, which was able to bloom and prosper. During its peak, Rome's influence reached parts of England, Africa, Syria and Spain. Roman roads stretched some 50,000 miles and Roman law ruled one in every four people.[21]

The influence of these two former superpowers is still evident today. The Roman Empire provided us with some of the most developed opportunities of representative democracy (though it differs greatly from what we know today). And the Chinese Dynasty paved the way for the spread of religion and culture from the East.

However, despite these successes, both empires eroded. Their great powers for hundreds of years ended and gave way to a new order in the world.

As we look at the past, it is evident that once highly regarded giants are susceptible to a rapid decline in power, strength and influence despite significant economic growth and success. Today, the great Roman Empire and the long reign of Chinese dynasties are merely echoes of the past. Even though the seeds they planted in terms of cultural and governmental thought are significant, their reigns still ended.

---

21 The Roman Empire: In The First Century, "The Roman Empire," PBS, http://www.pbs.org/empires/romans/empire/index.html.

No one would have imagined such a change in the political landscape back then, but we now know how quickly the world can change. We must be cognizant of change and the possibility that world powers can quickly rise and fall. We must also recall that we cannot assume their demise too early. The United States is struggling, for example, but the nation has shown throughout its history that it knows how to respond during tough economic times. This was true when the country was a developing industrialized nation during the Panic of 1819, the Depression of 1837, and the Depression of 1929.

But, given the lessons of history, it is unreasonable to think that today's global leaders will determine the political landscape of the future. With the re-emerging powers in the East, it is clear that the competition to influence the world's global economic development is fierce.

Twenty-first century equivalents of Silk Roads are opening up every day. As they open, they are redefining the world economy. One signal of an economic power shift from the West to the East came from HSBC, the bank founded in Hong Kong and Shanghai in 1865, which moved its headquarters to London in 1993. The bank recently announced it is moving the office of the Group Chief Executive back to Hong Kong to focus on emerging Asian markets. For the largest non-government-backed bank in the world to make such a move, marks a recognition of the shift in global economic power from the West to a re-emergent East.

Newsweek suggests that two great shifts in power have occurred over the past 400 years, including the rise of Europe and the rise of the United States. It concludes, "China's rise, along with that of India and the continuing weight of Japan, represents the third great shift in global power – the rise of Asia."[22]

---

22 Fareed Zakaria, "Does the Future Belong to China?," *Newsweek*, May 9, 2005, http://www.newsweek.com/id/51964.

Evolution of American Leadership

NOTABLE U.S. MINORITY FIRSTS
First African American Senator
Hiram R. Revels *(1870-1871)*
First Asian American Senator
Hiram L. Fong *(1959-1977)*
First Hispanic American Senator
Octaviano Larrazolo (*1928-1929*)
First American Indian Senator
Charles Curtis *(1907-1913;*
also was U.S. Vice President 1929-1933)
First African American Supreme Court Justice
Thurgood Marshall *(1967-1991)*
First Woman Supreme Court Justice
Sandra Day O'Connor *(1981-2008)*
First African American Secretary of State
Colin Powell *(2001-2005)*
First Female African American National Security Adviser
Condoleezza Rice *(2001-2005)*
First African American Secretary of Education
Rod Paige *(2001-2005)*
First Female Asian American Secretary of Labor
Elaine Chao *(2001-2009)*
First Female African American Secretary of State
Condoleezza Rice (2005-2009)
First Hispanic American Attorney General
Alberto Gonzales *(2005-2007)*
First African-American President
Barack Obama *(2009-)*
First Hispanic Supreme Court Justice
Sonia Sotomayor *(2009-)*
Source www.carnellknowledge.com

We need not look far into U.S. history to see that changes are occurring in all facets of our lives, including our own President. The election of President Barack Obama is yet another indication of the notion of change driven by the people. His election represented a major shift in the U.S. geopolitical landscape. Just 40 years earlier, in major parts of the country, black Americans could not vote, were required to ride in the back of the bus, and were relegated to separate and unequal classrooms. The election of 2008 was a stunning reminder of how far the U.S. has come culturally on racial issues.

Although this is significant change, it has come via evolution rather than revolution. However, even in this context, it seems that some of the country's indiscretions are in the not so distant past by comparison. The United States, often viewed as the trailblazer of civil rights, sometimes finds itself lingering behind "less advanced" societies that have grown in their own right.

For example, it was in 2008 that the United States had its most serious female presidential candidate. Meanwhile, countries less developed, economically speaking, were electing females to the highest offices nearly 40 years ago.

In 1974, when Argentina elected Maria Estella Martínez de Perón as President, the United States Supreme Court was just three years removed from ending hiring discrimination based on gender. Even in India, where traditions like dowries for women still exist, Pratibha Patil was elected president in 2007 by two-thirds of the votes.

The great triumphs made in government by the African American community are perhaps even more significant. It was just over 40 years ago that the Voting Rights Act was passed. This act prohibited the use of voter qualification tests that discriminated against many of the country's minorities. This was the beginning of an end to a tumultuous period in American history. In 1989, Colin Powell became the first African American to serve as Chairman of the Joint Chiefs of Staff. And in 2008, America elected its first African American President of the United States.

Because of term limits, America's political leadership changes every four to eight years. And usually with it, the economic course of the country changes as well. However, it is not politics so much as the financial climate that normally dictates economic changes. Therefore, there are constant threats to the established economies.

As I travel around the world, many non-Americans feel compelled to share their views of the ills of our country. I sometimes relay the very open dialog we have within the U.S. about how to improve – a discussion and debate that is never ending and never short of suggestions. It is not news that there are criticisms of capitalism. Even our founding fathers could not agree on its merits. Thomas Jefferson, some argue, was never a big fan of capitalism. He opposed the burgeoning control of a centralized government subjugated to big capital, and was not in favor of a central bank. He preferred to tout a decentralized form of government. In addition, "The idea of making profit," Jefferson argued, "was unethical."[23]

We should not discount the attributes that led to the adoption of American free market enterprise and to expanded civil liberties. While there is always debate of whether these go too far or not far enough, this sense of rugged individualism has been the basis of continued evolution for any culture. This includes the great melting pot society that is alive and well today in America.

The United States continues to press on and demonstrate its capacity to provide many opportunities for all people. At the same time, the nation continues to evolve and further its growth while competing with geographies that have a much longer history of sophisticated civilizations.

A Conflicted World

A cursory glance at the existing global conflicts would have many worried that the world is coming to end. Or, perhaps, that the rapid acceleration of human population has caused the inevitable conflicts. Or that religion and intolerance has led to the imminent downward cycle. Or, perhaps the devolution of the human race and its increasing focus on personal needs above the larger social needs has placed us teetering on the brink of a long, steady, downward cycle. The theories are virtually endless. Nevertheless, these prognostications

---

23 Brennan McKinney, "Thomas Jefferson and Agricultural America," *Associated Content*, December 15, 2008, www.associatedcontent.com/article/1283884/thomas_jefferson_and_agricultural_america.html?cat=37.

have existed for generations, just as there have always been conflicts around the world. This is important to note because most significant economic and cultural growth has happened in times of stability.

These invariable conflicts afflict people all over the world. The constant turmoil inhibits economic and cultural manifestation as well as shapes the basic expectations of people in certain parts of the world.

While nuclear threats, religious based factions and other conflicts create true global instability, it is the social dynamic and human governor of how we should live together and exist that ultimately sets the social standard by which we all decide to live. Religious beliefs, laws and governments help people create a lifestyle that is tolerable and creates the basis for growth. Often historical challenges are ultimately stabilized by the social desire of the people to live a fulfilled life, find meaning, be safe, care for our young, and have a chance to lead the life we desire. Of course, war and conflicts have existed as long as people have walked the earth. Even in Mesopotamia, documentation exists of significant war and conflict. This area between present-day Iraq and eastern Syria dates back to 6000 B.C. and is one of the first civilizations in history. The Greeks called the area Mesopotamia, literally the "land between the two rivers," a reference to the Tigris and Euphrates basin.[24] While this civilization gave the world its first system of writing and the world's earliest surviving work of literature, it was also known for constant chaos, almost the antithesis of what stability brings in today's world. Nevertheless, the Mesopotamian civilization was far more developed socially and technologically than most other cultures at the time. Their advanced laws, military

24 Richard A. Gabriel and Karen S. Metz, *A Short History Of War: The Evolution of Warfare and Weapons* (U.S. Army War College: Strategic Studies Institute: 1992), http://www.au.af.mil/au/awc/awcgate/gabrmetz/gabr003c.htm.

organization and technology gave the region a better measure of stability, which allowed them to expand development.

One benefit, some argue, for the U.S. is that for well over 100 years, no wars have been fought on American soil. This has resulted in relatively long periods of homeland security, allowing the opportunity for greater cultural expression and financial stability. In fact, we enjoyed remarkable stability until that fateful day on September 11, 2001, when acts of terrorism devastated the nation, triggered two wars and dramatically changed travel, communications, how we transfer money, and our feelings of personal security.

Often, periods of stability can follow periods of instability; this tension creates change, which can lead to growth. Once stability is established following the current crisis, it may result in a refreshed outlook to the underlying issues, which may lead to rapid expansion. At the same time, other nations are now growing more rapidly and industrializing due to extended periods of political and economic stability.

Nevertheless, whether by a will of the people or a philosophy of individualism, a culture exists in the U.S. that adapts to change and continues to move forward in spite of the many predictions of its imminent demise. The cynics are louder again and more clearly aimed at the industrialized nations that have contributed so handsomely to this current economic environment. Many point to this as evidence that the great capitalistic experiment has failed us completely. On the other hand, capitalism's characteristics allow for a constantly shifting playing field, which in turn demands that only the most innovative, creative, and flexible are able to navigate an unending stream of change and challenges.

Time will determine if these cultural traits are enough to compete in a world that is changing and where many of the historical advantages no longer exist.

Shifting Demographics

While the economic, geopolitical and leadership changes impact our human landscape, so too does our rapidly changing demographic profile. Because demographics are also changing in fundamental ways, they could have a profound effect on human behaviors.

The significance of today's trends can greatly impact the way people interact with their financial institutions as well. As the industry addresses other changes in our external environment, they must also pay attention to the shifting profile and needs of the human population. How they navigate forward will depend in part on mapping their strengths with these changing human attributes.

While it seems obvious that Generation X (born 1965 - 1978) differs greatly in behavior and attitude from its younger counterpart, the Millennials, there are still common threads that tie each generation to the next. While Millennials prefer to work in a collaborative environment, Generation Xers are highly individual workers. Yet both focus on and value a balance of work and life.[25]

Not only do gender and age based trends create differences, the interconnectivity of generations has an impact as well. It is also important to view these generational differences on an international level. In an article published in the Harvard Business Review, Tammy Erickson compares the models from India and the United States for varying generations. Erickson concludes that recent generations are more connected than ever because of technology's reach. As a result, the generational differences between the United States and India are

25 Denise Kersten, "Today's generations face new communication gaps," *USA Today,* November 15, 2002, http://www.usatoday.com/money/jobcenter/workplace/communication/2002-11-15-communication-gap_x.htm.

less significant for Generation Y than they are for the Traditionalists (born 1922 – 1945).[26]

While the global similarities will increase with younger generations, many industrialized nations are aging rapidly. In industrialized nations, 25 to 35 percent of the population is now over 65. In the United States, the 65 and older age group is the fastest growing segment of our population.[27]

This increase in seniors is in part due to changes in life expectancy, which has increased dramatically from only a few decades ago. In 1970, the average life expectancy was 70.8 years for a person living in the United States.[28] In 2005, this number has increased to 77.8 years and is projected to continue to rise at least through 2020.[29]

In developing economies, the workforce will increase to 700 million by 2010. This is more than the total workforce in fully developed nations. Although the world's population growth has been decelerating since 1970, the growth in developing and emerging countries has offset the decline in growth rate among the more developed countries.

At the same time, America still attracts immigrants that want a chance at success. They make up a greater proportion of the total population than ever before, estimated at 15 to 20 percent in the most recent figures.

---

26 Tammy Erickson, comment on "Generational Differences Between India and the U.S.," Harvard Business Review Blog, comment posted February 28, 2009, http://blogs.hbr.org/erickson/2009/02/global_generations_focus_on_in.html.

27 Jean-Pierre Lehmann, "Developing Economies and the Demographic and Democratic Imperatives Of Globalization," *International Affairs*, (2001).

28 Centers for Disease Control and Prevention, *National Vital Statistics* Report, Volume 56, Number 10, http://www.cdc.gov/nchs/data/nvsr/nvsr56/nvsr56_10.pdf.

29 U.S. Census Bureau, *Expectation of Life and Expected Deaths by Race, Sex, and Age*, http://www.census.gov/compendia/statab/cats/births_deaths_marriages_divorces/life_expectancy.html.

Other significant highlights gleaned from U.S. Census figures include:

- The percentage of the population in the "working ages" of 18 to 64 is projected to decline from 63 percent in 2008 to 57 percent in 2050.
- The working-age population is projected to become more than 50 percent minority in 2039 and be 55 percent minority in 2050 (up from 34 percent in 2008).
- In 2030, when all of the Baby Boomers (born 1946 – 1964) will be 65 and older, nearly one in five U.S. residents will be 65 and older. This age group projects to increase to 88.5 million in 2050, more than doubling the number in 2008 (38.7 million).
- Minorities, now roughly one-third of the U.S. population, are expected to become the majority in 2042, with the nation projected to be 54 percent minority in 2050. By 2023, minorities will comprise more than half of all children.
- The Hispanic population is projected to nearly triple, from 46.7 million to 132.8 million during the 2008 to 2050 period. The Asian population projects to climb from 15.5 million to 40.6 million, rising from 5.1 percent to 9.2 percent.

It is clear we must deal with this latest immigration influx as we have throughout our history. Those who successfully adjust to the changing landscape will be positioned to win.

While the female population has made significant strides, the perception of gender roles is changing slowly. In addition, their significant gains in the workforce have created other challenges, as women are still the principal caregivers of children and caretakers of the home.

Among Generation Y, women are just as likely as men to want jobs with greater responsibility. This was not the case among employees under age 29 as recently as a decade and a half ago.

Women's labor force participation has increased significantly in recent years. Women are gaining on men to the point that they shortly will outnumber men in the workforce, stemming from changes in women's roles and men experiencing enormous job layoffs during the recession.[30]

In addition, the most recent data from the Bureau of Labor Statistics indicate that unemployment rates have increased more rapidly for men than for women over the past year or more.

The bottom line is that proportionately, employed men and women are approaching parity, and women may already represent a greater share of the wage and salaried labor force than men.

Women's level of education has also increased relative to men, accounting for the majority of bachelor's degrees since 1982 and more master's degrees than men since 1981.

Women earned 58 percent of all bachelor's degrees and 60 percent of master's degrees in the 2005-2006 academic year (the most recent year for which data are available). By 2016, women appear heading to earning 60 percent of bachelor's, 63 percent of master's, and 54 percent of doctorate and professional degrees. [31]

The gender gap in earnings is slowly narrowing. However, women still make less money than men, primarily because they tend to go

30 Cathy Arnst, "Will The Recession Change Gender Roles?," *Business Week*, February 6, 2009, http://www.businessweek.com/careers/workingparents/blog/archives/2009/02/will_the_recess.html.

31 Ellen Galinsky et al., "2008 National Study of the Changing Workforce," Families and Work Institute, http://www.familiesandwork.org/site/research/reports/Times_Are_Changing.pdf.

into careers that pay less, such as teaching instead of engineering, or work in part-time positions so they can care for their children. Today, women hold 49.1 percent of non-farm jobs, but they work fewer total hours than men and are more likely to work part-time jobs without health benefits.[32]

While narrowing, the pay gap still exists. In 1979, the average full-time employed woman earned 62 percent of what men earned weekly. By 2007, however, the average full-time employed woman earned 80 percent of what men earned weekly, an impressive increase, but still short of their male counterparts.[33]

Higher levels of education increase women's earnings, just as they do for men.[34] Other findings show traditional gender roles have lost favor between both sexes. Most men and women disagree that men should earn the money and women should handle childcare.

Demographics are also changing in a profound way when it comes to religion. When it comes to worship preferences, the United States is both multi-faith based and growing as non-faith based in some respects. According to the American Religious Identification Survey (ARIS) 2008 survey, most religious groups in the U.S. have lost ground. For example, the percentage of people who call themselves Christian in some way has fallen by more than 10 percent in a generation with 86 percent of American adults identified as Christians in 1990 and 76 percent in 2008.[35] Clearly, Christians represent a substantial majority in the U.S.

---

32 Arnst, "Will The Recession Change Gender Roles?", 2009.

33 Galinsky, "2008 National Study of the Changing Workforce."

34 Sharon Jayson, "Gender Roles See a 'Conflict' Shift In Work-Life Balance," *USA Today*. Updated March 26, 2009, http://www.usatoday.com/news/health/2009-03-26-work-life-balance_N.htm.

35 Cathy Lynn Grossman, "Most Religious Groups In USA Have Lost Ground, Survey Finds," *USA Today*, March 27, 2009, http://www.usatoday.com/news/religion/2009-03-09-american-religion-ARIS_N.htm.

Other religious affiliation trends are also taking place in the U.S. today. The U.S. Religious Landscape survey comments, "More than one-quarter of American adults (28 percent) have left the faith in which they were raised in favor of another religion – or no religion at all." Additionally, the study reports, "While those Americans who are unaffiliated with any particular religion have seen the greatest growth in numbers as a result of changes in affiliation, Catholicism has experienced the greatest net losses as a result of affiliation changes."[36]

The 2008 ARIS study also revealed that so many Americans claim "none" as a religion (15 percent, up from 8 percent in the 1990 ARIS survey), that this category now outranks every other major U.S. religious group except Catholics and Baptists. "The challenge to Christianity ... does not come from other religions but from a rejection of all forms of organized religion," the report concludes.

In comparison, the global population is much different. A breakdown of the religious affiliations globally follows: Christians 33.32 percent (of which Roman Catholics are 16.99 percent, Protestants 5.78 percent, Orthodox 3.53 percent, and Anglicans 1.25 percent), Muslims 21.01 percent, Hindus 13.26 percent, Buddhists 5.84 percent, Sikhs 0.35 percent, Jews 0.23 percent, Baha'is 0.12 percent, other religions 11.78 percent, non-religious 11.77 percent, and atheists 2.32 percent.[37] In 2000, 32.88 percent of the global population considered themselves to be of the Christian faith, which shows that Christianity may be spreading globally.

From the beginning of mankind, there seems to have been a desire for meaning and a spiritual existence. It appears that desire remains among the majority of Americans. This search for purpose

36 The Pew Forum On Religion & Public Life, "U.S. Religious Landscape Survey, Religious Affiliation: Diverse and Dynamic," February 2008.

37 Central Intelligence Agency, *The World Factbook*, https://www.cia.gov/library/publications/the-world-factbook/fields/2122.html.

and meaning can manifest itself in scientific pursuits, psychology interests, or religious beliefs. Some follow the traditional religions, others embrace paranormal beliefs, and many follow a variety of spiritual paths. "More than ever before, people are just making up their own stories of who they are. They say, 'I'm everything. I'm nothing. I believe in myself,'" says Barry Kosmin, ARIS survey co-author.[38]

Whether the trends are gender, racial, generational or religious, the demographic shifts impact how we live and interact. Understanding the current trends and implications will be important to connecting with others and better serving entire communities.

Technology and the Knowledge Worker

Science and technology are shifting at an enormous rate and affecting how we interact, communicate, live, socialize and consume products and services. At the same time, the shift to a knowledge economy where the knowledge workers differentiate themselves in innovation, thought leadership and efficiency, is also impacting how we compete for a leadership role in the global economy. This next phase of global competition is putting pressure on education and the cooperation between governments, business and the educational systems to produce the most competitive labor forces for the knowledge economy.

The stem-cell issue is one example. While the religious and political ramifications of stem-cell research are debated in the United States, other nations are collaborating with their government, business and education groups in a variety of areas to provide funds, research facilities and attract the top scientists around the world.

---

38 Grossman, "Most Religious Groups In USA Have Lost Ground, Survey Finds," 2009.

China, the United Kingdom and Singapore all committed significant capital for stem-cell research and have projects well underway. In the United Kingdom, manufacturing facilities for pure stem-cell production for research were approved. China has attracted some of the best scientists with funding and facilities where discoveries are already being made.

The Singapore government is funding a biotech complex where scientists working for government-funded, private biotechnology or pharmaceutical companies can collaborate.[39]

Despite this progress in other nations, the RAND Corporation's U.S. Competitiveness in Science and Technology report reveals American science is holding its own, for now. According to the study, the United States accounts for 40 percent of total world research and development spending in the biotech field and 38 percent of patented new technology inventions by the industrialized nations of the Organization for Economic Cooperation and Development (OECD).[40]

Key findings in the U.S. BioScience Industry Report from 2008 show that there is a growing employment base in American biosciences up from 1.2 million in 2004 to 1.3 million in 2006. In addition, academic bioscience accounted for 60 percent of the U.S. R&D dollars spent, totaling $29 billion in 2006.[41]

---

39 Terri Somers, "Worlds Apart," *Union-Tribune*, December 17, 2006, http://archives.signonsandiego.com/news/business/biotech/20061217-9999-lz1n17somers.html.

40 Titus Galama and James Hosek, "U.S. Competitiveness in Science and Technology," RAND National Defense Research Institute, (2008), http://www.rand.org/pubs/monographs/2008/RAND_MG674.pdf.

41 Battelle Technology Partnership Practice, "State bioscience Initiatives 2008," http://www.bio.org/local/battelle2008/State_Bioscience_Initiatives_2008.pdf.

How sustainable this will be, is a major question. *Newsweek* reported that at the 2005 Science and Engineering Fair, six million pre-college students from China competed as compared to only 65,000 pre-college American students.[42] More than 200 million children in China study English (while less than 50,000 elementary and secondary school children in the U.S. study Chinese), according to statistics supplied by the U.S. Department of Education's No Child Left Behind program. This means that soon China will be the number one English speaking country in the world, at a population rate four times the size of the United States.

Although the West pioneered the information technology industry, we are now leveling off, or possibly falling behind, our international counterparts. We continue to be the innovators in the industry, but we no longer control it.

The U.S. needs people skilled for a knowledge-based economy.[43] In a white paper released by The Information Technology Association of America, it states, "Without disciplined, purposeful action now, the nation's high technology future, and therefore its economic future, are at risk." To remain globally competitive, America must double the number of Science, Technology, Engineering and Math graduates over the next 10 years, from approximately 430,000 to 860,000.[44]

The post-industrialization of the global economy has produced the "knowledge economy" where the playing field has changed as investments in educational systems and access to raw materials changed. Global alliances surrounding production capacity, raw materials, military strength and an emerging global cultural shift

---

42 Zakaria, "Does the Future Belong to China?" 2005.

43 Lehmann, "Developing Economies and the Demographic and Democratic Imperatives Of Globalization," (2001).

44 Innovation and a Competitive U.S. Economy: The Case for Doubling the Number of STEM Graduates by The Information Technology Association of America, 2005

towards innovation as a standard for differentiation has eroded the global positioning of established economies and caused a refocus on how to compete and win. With the global economic collapse as the backdrop, these points of clarity have emerged as critical concepts to grasp in a powerful new world order, potentially setting the table for the future economic vitality of nations and entire geographic regions.

Macro trends impact almost every aspect of our lives, punctuated by the recent financial crisis around the world. Some of the trends have longer-term implications and may have contributed to the size and scope of the economic downturn. They represent a manifestation of larger shifts in the geopolitical, educational and globalization areas. The financial services industry, being a reflection of so many aspects of what is happing in communities around the world, is forced to deal with these trends both large and small. Research validates what many already knew, which is that the rate of change in almost every aspect of our lives is accelerating.

As we see the early signs of stabilization that could lead to the end of the Great Recession, we are left to reflect on what have been the most dramatic changes to our economic landscape. The human race has overcome significant changes in the past and, with history as a guide, we will no doubt find a way to the new tomorrow.

Staying ahead, or at least keeping pace, with our changing world is increasingly complicated. Sea changes and subtle shifts in so many facets of the lives of people we serve challenge how we sell and market. Changes to financial systems, education, demographics, geopolitics and other areas affecting the global monetary systems require those who impact the economic engines to become more aware of the impact of these changes.

The accelerated rate of change will require a more flexible environment where static plans over multiple years will be replaced by a more agile operating environment designed to navigate the fog of the future better, guided by principles and values. This may be one of the most important strategic points for financial institutions to recognize as they move forward. With this as a backdrop, we'll turn to the specific changes that impact the financial services market and discuss how the fundamental business model is under severe pressure and how to approach the future of our industry.

2: Banking Industry Undergoes Fundamental Change

"We can't solve problems by using the same kind of thinking we used when we created them." - Albert Einstein

---

Can you believe what has happened to the financial services industry? Venerable, familiar, established organizations dissolved or altered forever. Bear Stearns, 85 years old, $350 billion in assets; AIG, 89 years old, $1 trillion in assets; Merrill Lynch, 94 years old, $1.6 trillion in assets; and Lehman Bros., 158 years old, $639.4 billion in assets; all swept away or salvaged in the aftermath of the subprime mortgage crisis.

Although seemingly taking place in the blink of an eye, this financial meltdown did not occur overnight. Before the profusion of inexpensive credit, government policies promoting home ownership, and an over-exuberant housing market, which precipitated many of the changes that eventually leaked into all areas of the global economic engine, the fundamental economic model for financial institutions was already undergoing intense strain.

The resulting Great Recession has had a genuine social and global effect on how humans interact with their financial institutions. It has tested the trusted foundation of financial institutions, brought into question the cost and benefits of government oversight - which many believe failed miserably - and created an active debate about the future of the industry. It has also prompted a review of how the financial industry views risks and manages global exposure.

Although the tempo and implications of global socio-economic, demographic and other changes discussed earlier are having a profound impact on how we live and interrelate, the financial industry has been experiencing significant changes for a number of years. Prior to the dramatic recent events of the global financial institution-led economic slowdown, the industry's basic business model was under attack. The surge just prior to the subprime crisis could not hide the underlying trends that were pressuring the industry. This shift, as it accelerates, should keep boards of directors and senior executives of financial institutions sleepless as they emerge from the hangover of the cheap credit party and begin to realize the problems have not gone away with mere time. They are being forced to take their medicine of rethinking how they will compete and focus on the value they add to the market.

In today's marketplace, powerful forces such as continued margin compression, channel proliferation, real time demands by account holders, intensifying regulatory changes, global technology shifts, improved accessibility to comparative shopping for like products, and fraud and security risks have all converged to accentuate what is already an intensifying competitive financial services landscape.

While financial institutions, in particular smaller and mid-sized institutions, have been able to leverage their long-standing competitive attributes of personalized service, trust and community affinity, the industry is struggling to deal with the additional pressures that come from a dynamic and ever-changing business environment.

The Subprime Bubble

The first obvious question is why did financial institutions make all those bad loans? Lenders sanctioned many individuals that were previously seen as not creditworthy. They opened up the door and made obtaining home loans easier than at any time in history. Those who previously did not qualify for a mortgage now suddenly found it relatively easy to receive one.

For most of the twentieth century, mortgage lending took place primarily at banks, thrifts, credit unions, and savings and loan associations. The most common type of mortgage was a fixed-rate mortgage and most of the financial institutions originating mortgages held and administered the mortgages that they originated on their own books.

As the 1960s were winding down and the 1970s began, the U.S. Congress, in part to make home ownership more affordable for a greater number of people, chartered Fannie Mae and Freddie Mac as government-sponsored enterprises (GSE).[45] Among their contributions to the housing and home mortgage marketplace, Fannie Mae and Freddie Mac created a liquid secondary market for mortgages that had previously been held in-house by financial institutions. This created a vehicle for other investors to participate in these mortgages by allowing financial institutions to sell them into a secondary market, which then created new investment products and spawned a creative market for packaging and pricing these new instruments. At the same time, the GSEs allowed financial institutions to sell these loans immediately after origination. This created a high fee income transaction without the credit risk on their books and freed up capital to make additional loans. Mortgage originators blossomed as they incentivized sales people to put these new classes of loans into the market for eager potential homeowners who might otherwise be shut out of the housing market due to lack of creditworthiness or the cost of the home they wanted to buy. This was buttressed by political leadership, which had the desire to provide more access to mortgages for those who might not previously qualify for the political goal of increased homeownership. At the same time,

45 Fannie Mae was originally created in 1938, but until its privatization in 1968, it was a part of the U.S. government.

it was a key tool to alleviate the pain of the intensifying fundamental shifts in banking by providing relatively easy and profitable new revenues for the financial services industry.

This convenient new revenue stream offered many institutions an obvious path to increasing profits without facing the more fundamental shifts in their business model. Rather than sharpening their competitive differences, upgrading their technology tools, and becoming more flexible and efficient at serving their communities better, many large institutions pushed aggressively into what must have seemed like a gold mine. At the same time, it masked the deteriorations in the base business of banking.

The availability of cheap credit and the availability of creative subprime mortgages drew in speculators and led to a significant increase in housing prices across much of the country. Like most bubbles, when it burst, it had a dramatic impact on the economy. Because the investment vehicles that were backed by these mortgages were so prevalent in portfolios around the world, the impact was global. Many attributed the source of the economic slide to the financial institutions that pushed these mortgage loans. Still others blame the U.S. Congress and a policy that promoted increased home ownership without monitoring more closely what was happening with Fannie Mae and Freddie Mac.[46]

As a result, in the four years leading up to 2007, the U.S. subprime market had grown from $332 billion to $1.3 trillion, an increase of over 290 percent. The exposure to Fannie Mae and Freddie Mac, both directly and indirectly peaked at approximately $5.1 trillion, roughly half of the U.S. mortgage market. As losses began to mount, it was clear that the Federal Housing Finance Agency (FHFA) would have

---

46 Barry Nielsen, "Fannie Mae, Freddic Mac and The Credit Crisis of 2008," Investopedia, http://www.investopedia.com/articles/economics/08/fannie-mae-freddie-mac-credit-crisis.asp.

to act, as the extent of the losses could not be absorbed within the existing structure of Fannie Mae and Freddie Mac. By September 2008, the market believed the firms were in financial trouble, and the FHFA put the companies into "conservatorship."

Evidence of a bubble being formed is that in the same time period as the dramatic increase in subprime mortgages (2003 to 2007), government-backed single family mortgages actually decreased by 58 percent[47] while the U.S. population grew by only 3.7 percent and the increase in subprime mortgages was in excess of a staggering 290 percent.[48]

The overexposure to debt and credit guarantee providers as the market began to falter was so significant that a government bailout was required for financial institutions. This systematic risk was allowed to balloon, and unlike the largest banks in the world, no bailout money has been repaid.

In order to survive, a long list of well-known financial institutions faced difficult decisions by 2008. Financial institutions were dropping like dominoes in what was clearly not a table game. They were fighting for their own survival amid the most dangerous financial crisis since the Great Depression.

The subprime crisis was exacerbated by the rush for every additional basis point of earning. It ignited a cascade of attention-grabbing headline bombshells and a massive remapping of the financial landscape that involved some of the most powerful players in the marketplace. Nearly all major players received assistance under the

---

47 Forrest Pafenberg, "Single-Family Mortgages Originated and Outstanding: 1990 – 2004," Office of Federal Housing Enterprise Oversight, http://www.fhfa.gov/webfiles/1151/mortmarket1990to2004.pdf.

48 U.S Census Bureau, *Annual Estimates of the Population for the United States, Regions, States, and Puerto Rico*, http://www.census.gov/popest/states/tables/NST-EST2009-01.xls.

Troubled Assets Relief Program (TARP), a provision of the Emergency Economic Stabilization Act of 2008.

- ***Bank of America Purchased Countrywide Financial Corp.*** on January 11, 2008. Countrywide had $408 billion in mortgage originations in 2007, and a servicing portfolio of approximately $1.5 trillion with 9 million loans.[49] With the arrival of the subprime mortgage crisis and secondary mortgage market disruptions, Countrywide Financial's business was subject to major questions regarding its solvency and made the company a potential bankruptcy risk.[50] (Note: In December 2009 Bank of America repaid $45 billion in TARP funds, plus an additional $2.7 billion in dividends to the Treasury).

- ***JPMorgan Chase Acquired Bear Stearns Companies, Inc.*** on March 16, 2008. Bear Stearns once boasted that it survived the Wall Street Crash of 1929 without an employee lay-off. However, with the subprime mortgage breakdown, 2007 was a disastrous year for Bear Stearns. With continued weaknesses in the subprime mortgage market, two internal Bear hedge funds collapsed in the spring of 2007 and started a market panic. In December 2007, Bear Stearns announced its first loss in its eight-decade history, about $854 million. The firm also wrote down $1.9 billion of its holdings in mortgages and mortgage-based securities.[51] Lender confidence in Bear

---

49 David Ellis, "Countrywide rescue: $4 billion," *CNNMoney.com*, January 11, 2008, http://money.cnn.com/2008/01/11/news/companies/boa_countrywide/index.htm.

50 Bank of America, Presentations and Press Releases, http://phx.corporate-ir.net/phoenix.zhtml?c=71595&p=irol-presentations and http://newsroom.bankofamerica.com/index.php?s=43.

51 "Bear Stearns Cos Inc.," *The New York Times*, http://topics.nytimes.com/top/news/business/companies/bear_stearns_companies/index.html.

Stearns quickly disappeared until it finally vanished altogether. Bear Stearns reluctantly asked the New York Federal Reserve for help to fend off counterparty risk that would result from a forced liquidation. The New York Fed then approached JPMorgan Chase & Co. to provide a 28-day emergency loan to Bear Stearns in order to prevent the potential market crash that would result from Bear Stearns becoming insolvent. (Note: JPMorgan Chase repaid $25 billion in TARP funds, plus an additional $1.7 billion in dividends and warrants).

- ***Lehman Brothers Filed for Chapter 11 Bankruptcy*** on September 15, 2008. Lehman Brothers generated a considerable share of its revenue through the issuance of mortgage-backed and asset-backed securities. When mortgage default rates began to rise and the demand for these securities began to disappear, there were no buyers in the market. The company's board of directors decided to file for Chapter 11 bankruptcy protection, which ended the firm's 158-year history and left it with the dubious distinction of the largest bankruptcy filing on record. Lehman Brothers held more than $600 billion in assets.[52]

- ***JPMorgan Chase Acquired Washington Mutual Bank*** on September 25, 2008. In 2007, Washington Mutual began to suffer a number of financial losses stemming from the aggressive sales of Option Adjustable Rate Mortgages (Option ARMs). With the subprime mortgage situation in full crisis mode, the corporation started reorganizing its mortgage loan operations, closing loan offices, and reducing staff to improve its operational efficiency. By September 2008, Washington Mutual's share price fell as low as $2 (compared to $30 in September

52 Lehmann Brothers, Press Releases, http://www.lehman.com/.

2007). Concurrently, Washington Mutual experienced a substantial deposit run off ($16.7 billion in deposits in a ten-day span) via the Internet (electronic banking) and wire transfers. Accordingly, the Federal Reserve and the Treasury Department pressured Washington Mutual to find a buyer, as a takeover by the Federal Deposit Insurance Corporation would have severely drained the FDIC insurance fund.[53]

- ***Wells Fargo Merged with Wachovia*** on October 9, 2008. Wells Fargo submitted its application to the Federal Reserve to expedite approval of the merger between Wachovia Corporation and Wells Fargo. In 2001, First Union Corporation announced it would merge with Winston-Salem based Wachovia Corporation. As the storm clouds of the subprime mortgage crisis gathered, Wachovia's purchase of Golden West Financial in 2006 started to play havoc with its financial competence. With a poor balance sheet and questions of solvency surfacing, the Fed and the Secretary of the Treasury suggested that Wachovia, which was already in talks with Citigroup and Wells Fargo, put itself up for sale. Wells Fargo became a more desirable suitor because it did not require financial assistance from any government agency.[54] (Note: Wells Fargo has repaid $25 billion in TARP funds, plus $1.4 billion in dividends).

- ***Bank of America Acquired Merrill Lynch*** on September 15, 2008. With the burst of the technology bubble in the early 2000s, Merrill Lynch scaled back its international

53 Eric Dash and Andrew Ross Sorkin, "Government Seizes WaMu and Sells Some Assets," *The New York Times*, September 25, 2008, http://www.nytimes.com/2008/09/26/business/26wamu.html?_r=1.

54 Wells Fargo, Press Releases, https://www.wellsfargo.com/press/.

operations and reduced its headcount dramatically. As it aggressively invested in collateralized debt obligations based on subprime mortgages, the U.S. subprime mortgage market collapsed. The situation forced Merrill Lynch to write-down and write- off nearly $8 billion in assets in the third quarter of 2007, and $16.7 billion in the fourth quarter, more than any other investment bank. In May of 2008, Merrill Lynch established a management group to determine how to divest itself of these CDOs (collateralized debt obligations) and other risky assets. According to this divestiture plan, the company sold $30.6 billion of CDOs for $0.22 on the dollar in July of 2008. After that sale, Merrill Lynch was incapable of shouldering subsequent losses and correspondingly sold itself to Bank of America.[55]

- ***AIG is Bailed Out*** on September 16, 2008. AIG suffered a liquidity crisis following the downgrade of its credit rating. Following the downgrade, the company was required to post collateral with its trading counter-parties, which led to AIG's liquidity disaster. To avert the company's collapse, the Federal Reserve announced the formation of a secured credit facility of up to $85 billion, secured by the assets of AIG subsidiaries, in exchange for warrants for a 79.9 percent equity stake and the right to suspend dividends to previously issued common and preferred stock. AIG accepted the terms of the Fed's rescue package and secured credit facility. This was the largest government bailout of a private company in U.S. history (this private bailout was smaller than the bailout of Fannie Mae and Freddie Mac a week earlier). On October 9, 2008,

55 Bank of America, Presentations and Press Releases, http://phx.corporate-ir.net/phoenix.zhtml?c=71595&p=irol-presentations%20and% 20http://newsroom.bankofamerica.com/index.php?s=43.

the company borrowed an additional $37.8 billion via a second secured asset credit facility created by the Federal Reserve Bank of New York. On November 10, 2008, the U.S. Treasury announced it would purchase $40 billion in newly issued AIG senior preferred stock through the TARP program. Further, the Federal Reserve Bank of New York announced that it would modify the September 16th secured credit facility (i.e., the Treasury investment would permit a reduction in its size from $85 billion to $60 billion).[56] As of this writing, $45.3 billion in TARP funding had been disbursed to AIG, with a total commitment of $69.8 billion. None of the TARP funding has been repaid.

- ***Goldman Sachs Became the Fourth Largest Bank Holding Company*** on September 21, 2008. The Goldman Sachs Group, Inc. received approval from the Federal Reserve to transition from an investment bank to a bank holding company. Since the spring of 2008, the company had been working with the Federal Reserve to review its liquidity and funding profile, capital adequacy, and overall risk management framework to substantiate its application for Bank Holding Company status. The new status allows the company to run commercial banking operations and gives its depositors insurance through Federal Deposit Insurance Corporation (FDIC). The deposits in turn allow Goldman Sachs to reduce its leverage ratio and hence reduce its risk of bankruptcy.[57] In November 2008, the organization received a New York bank charter. Gold-

56 Matthew Karnitschnig et al., "U.S. to Take Over AIG in $85 Billion Bailout; Central Banks Inject Cash as Credit Dries Up," *The Wall Street Journal*, September 16, 2008, http://online.wsj.com/article/SB122156561931242905.html..

57 Goldman Sachs, Press Releases, http://www2.goldmansachs.com/our-firm/press/press-releases/index.html.

man's move to obtain a state charter is a sign that the company may not want a consumer-oriented business that operates on a national level, but just wanted access to the Fed. (Note: In June 2009, Goldman Sachs repaid $10 billion to the TARP Fund, plus an additional $1.4 billion in dividends and warrants).

- ***Morgan Stanley Granted Federal Bank Holding Company Status*** on September 21, 2008. Morgan Stanley received approval from the Federal Reserve to transition from an investment bank to a bank holding company. According to Morgan Stanley, it sought this new status to provide it with the maximum flexibility and stability to pursue new business opportunities as the financial marketplace undergoes rapid and profound changes. In 2007, many of Morgan Stanley's mortgages and mortgage-backed securities saw their values plummet, which resulted in write-downs and write-offs totaling $10.8 billion in 2007, and a $3.5 billion quarterly net loss in the fourth quarter of 2007, the first in Morgan Stanley's history. Additionally, after the bankruptcy of Lehman Brothers and the sale of Merrill Lynch to Bank of America, Morgan Stanley's prospects as an independent firm became the subject of market analysts' speculation and led to talks of a merger or a buyout. Thus, the company sought Bank Holding Company status to sustain its existing business model.[58] (Note: Morgan Stanley has repaid $10 billion in TARP funding, plus an additional $1.3 billion in dividends and warrants).

In addition, there were large institutions in other countries with similar issues. Several British financial institutions came within

58 Morgan Stanley, Press Releases, http://www.morganstanley.com/about/press/index.html.

hours of a liquidity shortfall in the fall of 2008, as the U.K. financial system itself came to the threshold of disintegration.

Margin Compression Squeezes Financial Institutions

While the subprime crisis brought banking problems to the forefront, other issues and problems were bubbling to the surface before the breakdown, and are still threatening traditional financial institutions.

Operating margins have been in a slow steady decline for over a decade. The statistics are startling. Since 1994, the net interest margin at commercial banks has dropped by 113 basis points.[59] Over the same period, the net interest margin at savings institutions has decreased by 42 basis points,[60] while the net interest margin at credit unions has fallen by 75 basis points.

The future does not promise to get any better. While there may be short-term increases, I believe that the mounting threat of commoditization will lead to the industry's consumerization and usher in an era of unrelenting pressure on historical margin-related products.

The base business of buying and selling money will be under increased strain as the mystery erodes about how these products work. As information becomes more accessible and the consumer progressively more educated about banking, commoditization increasingly threatens the basic business. This is because individuals can now shop for financial products like any other consumer merchandise.

---

59 Federal Deposit Insurance Corporation, "Quarterly Banking Profile," March 20, 2009, http://www2.fdic.gov/qbp/2008dec/cb1.html.

60 Federal Deposit Insurance Corporation, "Quarterly Banking Profile," March 20, 2009, http://www2.fdic.gov/qbp/2008dec/sav1.html.

The resulting consumerization of banking will force financial institutions to review their strategic initiatives to better understand their market and product niches. Each financial institution will need to better define how it is distinctive and make that differentiation clear with every consumer interaction.

Regulatory changes have made it easier for other providers of financial products and services — insurers, brokerage firms, trust companies —to enter the market with traditional banking products and services as a way to diversify their own revenue streams and keep their clients in their family of offerings. These new players are often less concerned about the net-interest margins than they are about merely increasing wallet share so that they can cross-sell their traditional fee-based or premium based services to these new clients.

The result is even greater pressure on net-interest margins and on diversifying the product and service suite of traditional community-based financial institutions.

Even outside of the U.S., the financial industry is facing economic pressure. In the *Edmonton Journal*, CEO Garth Warner of Servus Credit Union, speaks about the institution's net income decline. Servus is the largest credit union in Alberta, Canada and faced a 28 percent drop from the same quarter in the previous year. "This is largely a result of margin compression due to low interest rates," states Warner.[61]

Australian institution, Bendigo and Adelaide Bank, dealt with similar difficulties. *The Australian* reported that the bank posted a 57.7 percent plunge in net profit in the second quarter of 2009 compared

---

61 Edmonton Journal, "Alberta's largest credit union reports drop in net income," *Calgary Herald*, September 30, 2009, http://www.calgaryherald.com/business/Alberta+largest+credit+union+reports+drop+income/2051386/story.html#.

to the same quarters in 2007 and 2008. The bank has raised capital to recover, but blames this significant loss to margin compression.[62]

It is clear that the playing field of the financial industry is changing dramatically due to these declining margins. New entrants into the banking market, for example, are less profitable than they used to be. Even new financial institutions are suffering. Due to thinner margins, de novos are taking longer to report quarterly profits and recoup losses.[63]

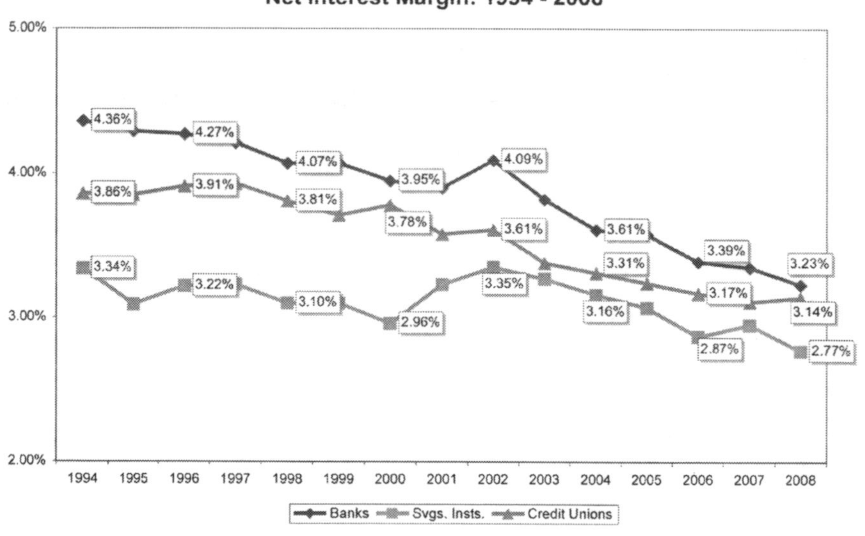

Net interest margins have declined significantly since the 1990s. The increasing commoditization of financial products is the primary factor behind this. This decline in margin income has been offset by growth in non-interest income, but this source of income is not expected to grow as rapidly going forward.

---

62 The Australian, "Bendigo and Adelaide Bank raising draws in $300m," *The Australian Business with The Wall Street Journal*, September 15, 2009, http://www.theaustralian.news.com.au/business/story/0,28124,26072766-643,00.html.

63 Bryant Rulz Switzky, "Stunted Growth," *Washington Business Journal*, July 24-30, 2009, https://www.accessnationalbank.com/home/fiFiles/static/documents/wbjarticle_july2009.pdf.

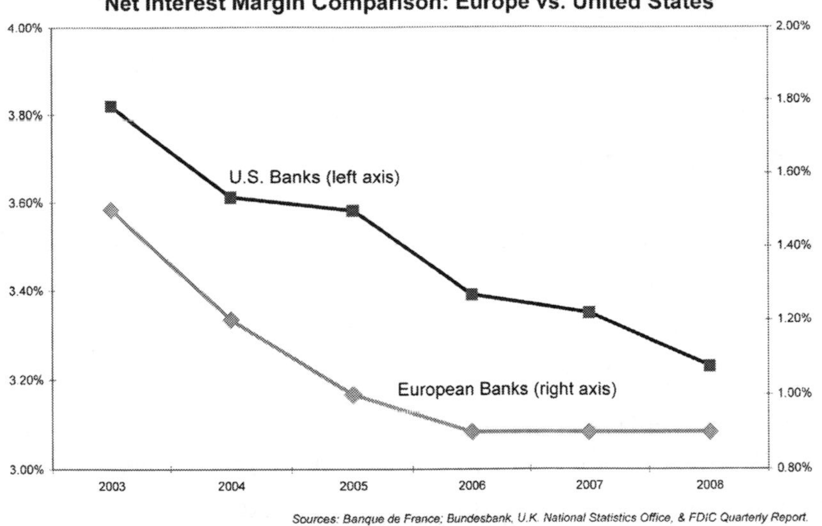

The trend in margin compression was not limited to the U.S.. European banks have also seen a significant decline in net interest margins.

Channel Proliferation

Congratulations, financial industry! You have succeeded in providing your clients with more ways to interact more frequently with their balances through more channels than ever before. The problem is that you are providing these additional conduits without receiving an increase in fee-income or assets. Therefore, it just cost you, the financial institution, more money to service the same balances. And the promise of fewer channels or capitalizing on improved demographics in the form of profits has yet to materialize.

While on the commercial banking side many of these services are chargeable, the retail account holder expects most of this convenience to be free. In addition, unless a financial institution has specific target demographics, it will likely have to continue to sustain multi-channel availability for the foreseeable future. The question is how to capitalize on the increased interaction with clients in a way that is meaningful to consumers and profitable to the institution.

The number of ways in which individuals interact with their financial institutions continues to expand. Institutions have discovered that, while the Internet is efficient as one of the lowest-cost delivery channels, it has not completely displaced the other means of communication, transactions and consumer interaction (e.g. ATM, branch, call center, voice, wireless device, etc.). In fact, clients are accessing their balances more frequently and in more different ways than ever before. Without an increase in assets or fee-income, this can become a net negative to a financial institution, which must bear the cost of providing these additional touch points. This channel proliferation phenomenon also illustrates the benefits of real-time processing, as it allows institutions to interact more powerfully with the individual by exploiting immediately updated information about the person.

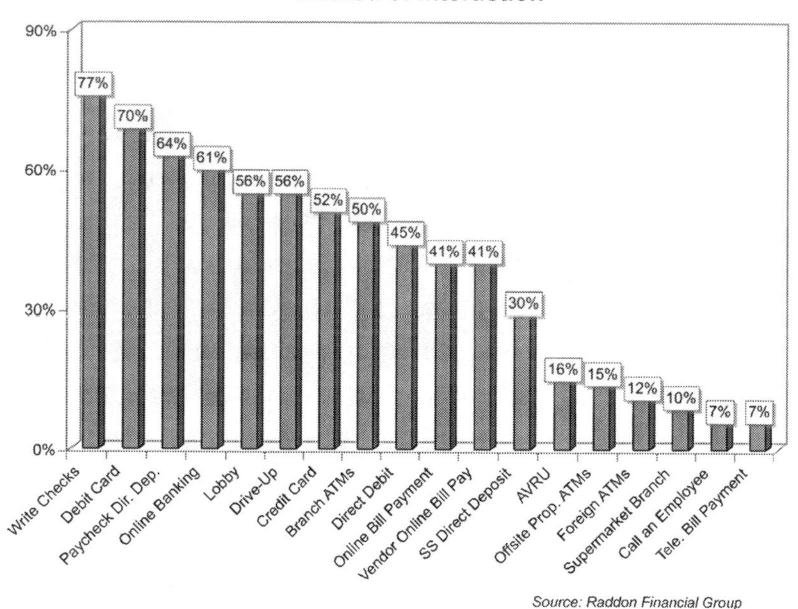

The proliferation of channel usage is clearly evident in this chart. Most channels tend to be complementary in nature. Introduction of a new channel typically doesn't eliminate other channels.

Industry analysts at Aite Group confirm that consumers are progressively more self-directed and are "better educated about their financial needs, more involved with their financial decisions, and less reliant on banks to tell them what they need." Aite Group analysts add "Consumers' online behavior hasn't displaced their offline behavior – it's added to it." [64]

Financial institutions must be able to manage all their channels efficiently and effectively, so that the increased level of interaction is not just consuming resources or adding expense. Instead, they must create an opportunity to cross-sell, reinforce service, and increase their wallet share on a per client basis.

Increased Regulatory Intensity

Another issue facing many institutions today is the rapidly increasing costs of regulatory compliance. The rate and significance of regulatory changes has created a scale litmus test that will threaten the viability of smaller financial institutions in the long run, if left unmanaged.

This comes not only from the rising cost of compliance, but is also a product of the highly inefficient regulatory and compliance apparatus. The current structure, for example, does not even regulate brokerage, insurance, and hedge funds where there can be substantial systematic risk. The multiple levels of regulation add fuel to the fire, and increase costs and inefficiency in the industry.

There is a complicated alphabet soup of federal and state administrative agencies, including the Office of the Comptroller of the Currency (OCC), the Federal Reserve Board (FRB), the Federal Deposit Insurance Corporation (FDIC), the Federal Trade Commission (FTC), the Internal Revenue Service (IRS), the National Credit

---

64 Ron Shevlin and Kate Monahan, "The Next Generation of CRM in Retail Banking: Sense-and-Response Marketing," *Aite Group*, (June 2009).

Union Administration (NCUA), and the Office of Thrift Supervision (OTS). In addition, there are reams of statutes, regulations, advisory opinions, and interpretive letters. Making this confusion of alphabets worse, is that it is not clear which of the agencies will survive.

New rules are being promulgated all the time. The frightening viewpoint for financial institutions is that there appears to be a rapidly increasing rate of these revised rules. The cost and complexity is also increasing, making compliance one of the most difficult challenges for small or medium-sized institutions. These regulatory implications will be troublesome for community-based institutions in particular, even though they did not participate at the level of larger institutions in the subprime market, pay exorbitant bonuses or base their income stream primarily on fee-income. However, they must often incur the costs of complying with changes aimed at these very issues even though they largely were not participating in them.

The events of 9/11 brought about additional regulation. It amplified the need for a change in the clearing system when grounded planes forced banks to take drastic alternate steps to ensure check shipments. It also resulted in the Patriot Act, which increased the ability of law enforcement agencies to search telephone, email communications, medical, financial, and other records.

Another change that impacted bank processing was Check 21. This statutory requirement created a new legal payment vehicle, a substitute check — a front and back copy of the original paper check — that banks can exchange electronically.

Also, new reporting requirements resulted from the U.S. Department of Treasury's $700 billion Troubled Asset Relief Program, that covered the financial system when it seemed on the brink of collapse. The OCC, for example, can request to review certain documents, policies, and other information related to institutions' luxury expenditure policy, as well as other compensation-related information.

Other additions include the Sarbanes-Oxley Act of 2002 (SOX), which requires significant enhancements to financial reporting systems of public companies, the Community Reinvestment Act (CRA), intended to encourage depository institutions to help meet the credit needs of the communities in which they operate and the Home Mortgage Disclosure Act (HMDA), which requires lending institutions to report public loan data. The list goes on and on.

The rate of changes, frequency, and sheer volume of regulatory demands on a financial institution continues to accelerate. This alone will cause some of the inevitable consolidations in the industry. Many institutions may soon discover what they unknowingly were not complying with; and determine that the cost of complying with new and changing regulations is no longer feasible, especially for some smaller financial intuitions. The increasing scale and intricacy of these obligations have led to a rapid growth of compliance costs, and at a rate considerably higher than that of revenues and profits.

A survey produced by the Deloitte Center for Banking Solutions showed compliance costs grew significantly faster than net income for financial institutions. As costs have risen, financial institutions appear to have responded by applying more human resources to monitor compliance versus technology resources to manage it.[65] This is a troubling trend for an industry that must find ways to become more efficient.

Compliance requirements drove increased capital spending on technology systems, software, and hardware. Spending on technology

65 Deloitte, "Navigating the Compliance Labyrinth: The Challenge for Banks," *Deloitte Center for Banking Solutions* (2007): 3.

systems was up more than 10 percent since 2002 at 90 percent of the institutions surveyed.[66]

While spending on technology vendors has also increased significantly, there is the lingering issue as to whether compliance processes could be made more efficient if more focus was placed on IT investments designed to reduce the costs of compliance compared to increasing human processes.

Executives estimated that their institutions' technology spending was 10 to 13 percent higher than it would have been without the additional compliance requirements they faced.

A NAFCU Flash report concludes that nearly half (43.1 percent) of the respondents expect to spend more money on compliance each year. Regarding which compliance areas would see this extra funding, respondents most cited technology (64.3 percent), followed by third-party support (53.6 percent) and internal staffing (39.3 percent). About two-thirds of the respondents said they have full-time staff dedicated to compliance issues (65.1 percent).[67] Additionally, IDC Financial Insights expects risk management and compliance spending to increase from $448.33 million in 2009 to $546.26 million in 2012 – an increase of 21.8 percent.

The primary responsibility of regulators is to enforce rules designed to strengthen the safety and soundness of our banking system. Unfortunately, one of the most important tools used by regulators and the industry to ensure that they are in compliance with the changing regulations and managing the financial position of the institutions

---

66 Deloitte, "Navigating the Compliance Labyrinth: The Challenge for Banks," (2007).

67 National Association of Federal Credit Unions, "CU compliance budgets unaffected by economy," February 18, 2009, http://www.nafcu.org/Template.cfm?Section=News&template=/contentManagement/contentDisplay.cfm&contentID=37819.

– the enterprise core processing system – contains some of the most outdated technology in use in any industry today. Yikes!

The irony is that often, regulatory policies themselves can mute the adoption of the latest technology, which makes the financial institutions less sound and less safe by adding layers of ineffective, disjointed and inefficient applications.

Fraud and Security

Our industry is by far the most targeted for abuse from fraud and theft. This has resulted in an increasingly sophisticated and costly effort that will continue to be a key component to risk management in the future.

There has never been a greater need for financial institutions to protect against security breaches than in this cyber world in which we live. An increasing threat from ever-sophisticated hackers and fraudsters as well as incursions on distributed networks and Web access, has put financial institution systems at more significant risk and challenged boards to diligently assess the possibility that security might be their Achilles' heel.

The statistics are sobering. According to the Privacy Rights Clearinghouse, data loss incidents involving more than 339 million individual records have occurred since January 2005.[68] The Identity Theft Resource Center of San Diego found that for 2009, the breach tally was close to 500 incidents with almost 60 of the incidents involving more than 27,000 reported records at financial institutions.

Confirming the existence of adequate safeguards is high on examiner watch lists today. Federal regulations and documents consistently address security. The U.S. government is stepping up requirements for

---

68 Privacy Rights Clearinghouse, *Chronology of Data Breaches*, http://www.privacyrights.org/ar/ChronDataBreaches.htm#Total.

financial institutions to identify and prevent fraud related incidents. In November 2007, the Federal Trade Commission and the federal bank regulatory agencies issued a rule entitled *Identity Theft Red Flags and Address Discrepancies Under the Fair and Accurate Transactions Act of 2003*, as directed by the Fair and Accurate Credit Transactions (FACT) Act. By the program deadline of November 2008, financial institutions were scheduled to provide for the identification, detection, and response to patterns, practices, or specific activities – known as "red flags" – that could indicate identity theft.

Failing to have adequate safeguards can be costly. Privacy and information management research firm the Ponemon Institute (Traverse City, Mich.) in its *2009 Annual Study: Cost of a Data Breach* reported that data-breach incidents cost companies $202 per compromised customer record in 2008, compared to $197 in 2007. However, "lost business opportunity," including losses associated with customer attrition and replacement acquisition, represented the most significant component of the cost increase.[69]

This spawns the debate about an individual's personal responsibility for data storage and protection in conjunction with the financial institution. It also highlights the need for a more integrated enterprise risk management system that monitors all forms of key enterprise risks, including fraud, control, credit and security related areas.

Financial services organizations increasingly understand the necessity to expand the extent of their security beyond its previous scope. Security touches all parts of an organization. It not only includes information security, but internal breaches, and physical security as well. Whether it is protection against fraud, privacy invasion, crippling viruses, identity theft, or a lone robber slipping a note to a teller, overseeing safety measures has become a 24/7 responsibility. This

69 Ponemon Institute, "Fourth Annual U.S. Cost of Data Breach Study," January 2009.

pressure to comprehend the risks is mounting as the consequence of highly publicized identity theft incidents and other security breaches, as well as legislation aimed at managing operational perils, becomes more public.

The easiest way to deal with a data breach is to prevent it from occurring in the first place. That is why the executive boards of emerging information and technology risk management programs now include chief executive officers, chief risk officers, IT chief risk officers, chief information security officers and, in many cases, a chief security officer.

Prevalent threats today are primarily electronic-based, and they are growing increasingly more dangerous. Social engineering attacks, such as phishing, are a chief concern because they are very difficult to prevent. These types of attacks prey on the trust account holders put in their financial institution. A bank or credit union can educate its clients, but the account holders themselves must be aware of what type of information they give out and to whom. The challenge is not only staying ahead of the fraudsters, but educating clientele on how to keep their data secure.

In addition, the criminals are clever and quick to evolve. When methods are discovered to protect account holders, the fraudsters find new methods to destabilize the system.

The new wave of security attacks has created a new lexicon of frightening terminology. Who would have thought that words like worms, phishing, spiders, reverse spider bombs, SMiShing, and vishing would evoke such fear in IT departments? Currently financial institutions have their hands full implementing intrusion prevention and detection for system attacks with these kinds of names.

Vishing, according to the FBI, is really just a new take on an old fraud — phishing. In a phishing attack, an account holder gets an email that claims to be from a financial institution or credit card company asking for an update on account information and passwords by clicking on a link to what appears to be a legitimate Web site.

Vishing schemes have variations on phishing that are somewhat different. In one version, an account holder gets the typical email, like a traditional phishing swindle. However, rather than being directed to an Internet site, the client is asked to call a number to provide the information over the phone. Those who call the "client service" number (a VoIP account, not a real financial institution) go through a series of voice-prompted menus that ask for account numbers, passwords, and other critical information. According to the FBI's Internet Crime Complaint Center (IC3), the number of "vishing" complaints received by the center is increasing at what it calls "an alarming rate."[70]

Another phishing variation targets mobile phone users. This new scam called SMiShing –phishing attack sent via SMS (Short Message Service) – entails someone receiving a message reading something like "We're confirming you've signed up for our XYZ service. You will be charged $2/day unless you cancel your order at our Web site." Many consumers, afraid of incurring premium rates on their cell phone bill, visit the Web site. Once they arrive, they are prompted to download a program, which is actually a Trojan horse.

It is no coincidence that SMiShing and mobile banking are both on the rise. Fraudsters are merely expanding into any vulnerable media. Criminals are always developing newer means of fraud.

---

70 Internet Crime Complaint Center, "Vishing Attacks Increase," January 17, 2008.

Some of the other implications of security issues could include:

- Denial of Service – Web sites, email, Internet applications, and other services become unavailable to the consumer. To the account holder – it looks as if your server is down which in turn threatens the trust individuals have in a financial institution's ability to secure their data.
- Trojans and Spyware – Web sites compromised by malware could endanger clients since access to applications, forms, files, and even just keystrokes can be easily copied and sent to a hacker when account holders access these services.
- Network Security – Misconfigured, outdated, or compromised firewalls allow attackers to jump from an external Web site to your production network. Most servers outside a "demilitarized zone" or production network do not have the same level of protection afforded to them as external Web services.

Financial institutions have to ask whether they are investing in the right tools. Certainly, they should have some basic protection. All financial institutions should have firewalls installed, intrusion detection for external and internal vigilance, anti-virus at every desktop, and some way to secure and monitor remote connections such as a virtual private network server (which some vendors use to support their software but it could also be an intrusion risk). However, simply throwing money at the security problem does not necessarily solve it; the challenge is to have the right balance of security integrated with internal processes. People also need to be proactive and understand what the risks are and then take action.

Channel proliferation, where clients are granted free access to their balances at the retail level, incurs more costs for an institution to maintain the same balances. It also incurs the cost of added fraud and security risks. This is just part of emerging increased costs tied to increased regulations, fraud prevention and security. It is an ideal area for global collaboration between institutions, vendors, governments and regulators. This silent war that rages around the world is costing hundreds of millions of dollars to defend. It also creates a tax on capitalism and the movement of money by those who endeavor to cause harm.

Industry Consolidation

Consolidation is a much talked about issue in banking that affects every size of financial institution in different ways. Many feel this compression is inevitable because of lower margins, scalability issues and industry fragmentation.

Smaller and mid-sized institutions emerged healthier as a group than larger institutions. They avoided many of the riskier investments and tended to have a narrower product and service suite. However, they lack the diversity, capital access, and the reach of larger institutions. The challenge has been for financial institutions to create infrastructures that exploit their size and provide better, more localized service.

Not to be ignored in the middle of the subprime crisis, the mega mergers, and stunning consolidations is that contraction among financial institutions has occurred steadily for years prior to the crisis.

From 2002 to 2008 there was shrinkage of 14.7 percent among banks, and a 22.1 percent drop in the number of credit unions. The expectation is that by 2012 there will be less than 7,000 credit unions in the U.S.[71] Canada also experienced major consolidation. The same holds true for almost all major developed nations around the world.

71 Federal Deposit Insurance Corporation, "FDIC Quarterly," http://www.fdic.gov/bank/analytical/quarterly/index.html.

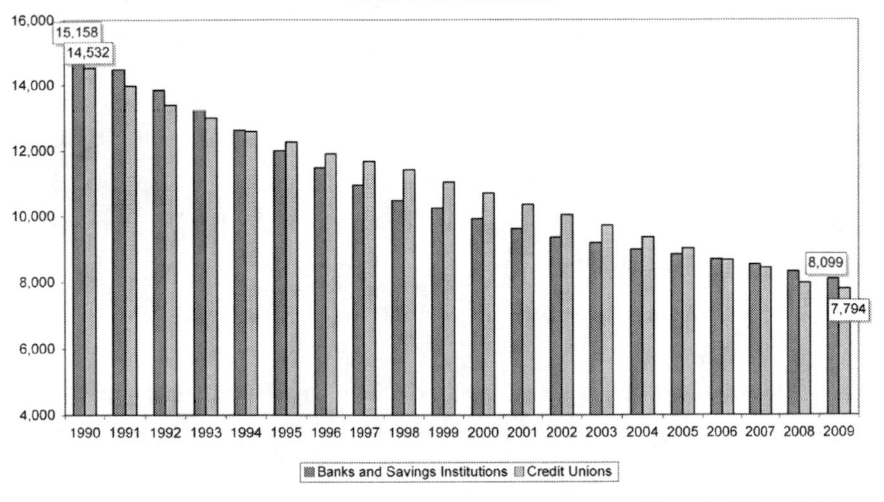

The number of banks and credit unions has been in decline for a number of years. Since 1990, the number of banks and savings institutions has declined by 46.6 percent, or 3.2 percent per year. The number of credit unions has declined by 46.4 percent, or 3.2 percent per year.

As seen in the graph above, since 1990, there are 46.6 percent fewer commercial banks and savings institutions, and 46.4 percent fewer credit unions. On average, both groups are declining by 3.2 percent per year.

In testimony regarding the subprime mortgage crisis to the U.S. House of Representatives, Alex J. Pollock, Resident Fellow at the American Enterprise Institute, stated, "In 1970, there were about 13,500 banks in the U.S. By 2005, there were about 7,500 – a reduction of almost half. The historical context is that the normal evolution of banking…to match the evolution to a national U.S. economy…was blocked for decades by artificial legislative and regulatory barriers… This resulted in a fragmented, less efficient, and more risky banking system composed of mostly undiversified small entities. When the barriers to the natural development were removed, the delayed consolidation, which is still in process, began."[72]

The consolidation, Pollock explained, has not reduced access to banks – just the reverse. In 1970, there were about 44,600 banking offices, mostly branches; by 2005, this had increased to about 80,300. Consequently, whereas the quantity of banks halved, the number of banking locations almost doubled, contrary to the popular 80's myth that bricks and mortar would disappear in the financial industry.

Consolidation is presumed to be needed because the industry's high operating and regulatory costs, combined with low margins, require scale and diversity.

However, as mentioned, smaller institutions performed better and took less risk during this most recent economic crisis. There also seems to be an ongoing demand for these community-based institutions that focus on service and community as evidenced by the continued launch of de novo institutions and recent market share gains by community banks and credit unions.

While some notably quiet large institutions weathered the storm without government intervention, it may be that the network of small and mid-sized institutions with their focus on local relationships, community investments, and development represented the strongest part of the economic engine during the recent financial industry crisis.

There has been a strong focus on institutions that are supposedly "Too Big to Fail." But perhaps it is the more vibrant, healthy, traditional community-based institutions we should focus on. A new mantra of "Too Small to Fail" may be more appropriate. This collaborative

---

72 Alex J. Pollock, "Bank Consolidation, Subprime Mortgage Issues, and the One-Page Mortgage Disclosure," *American Enterprise Institute for Public Policy Research*, May 21, 2007, http://www.aei.org/speech/26220.

network of community-based institutions focused on the basic needs of their communities, has demonstrated greater stability and more informed risk assessments, as they largely did not participate in the subprime and related areas. Also, because many hold the loans that they issued, they are much more aware of the risks and relationships that they are exposed to.

The industry, marketplace and regulators must find a way to allow greater flexibility and economic incentives for community-based, service-oriented institutions to survive and thrive.

Think Globally, Even When You Are Regional

No matter where you are geographically located, the globalization and increased interconnectivity of all people will have an impact on how our industry operates.

The interconnectivity of the world is greater than ever, electronically and socially. Today, there is pressure for all business to serve a more global marketplace. Even if your financial institution is not global, the degree of interconnectivity between all humans has an impact on how we serve account holders and their own view of what to expect from a financial institution.

It is increasingly difficult for institutions to operate in a comfortably controlled environment. Financial institutions have to consider relationships with other institutions, account holders, and other entities in their strategic planning. No matter where you are located, the flow and interconnectivity of information is so immediate between entities and individuals that financial institutions now must consider what is happening in other parts of the world, even if they are focused on winning locally in their marketplace.

In the process, financial institutions are part of connecting people, expanding language, culture, images and ideas along the way. On an economic level, technology provides opportunities for companies across the world to innovate through collaboration and exploit this increased interconnectivity.

Yet, as we have discussed, it appears unquestionably that one of the most crucial underlying bonds to each of these aspects of globalization is technology. In fact, according to Professor Robert Rycroft, globalization and technology have co-evolved together. He states that this is a result of "innovation networks," which he defines as "the complex webs of relationships among firms, universities, government agencies, and other organizations for generating and sharing knowledge relevant to technological innovation."[73]

Technological advances have continued to play a key role as an enabler of fulfilling human needs, including business strategies.

A few years ago, communication technology was still in its early phases and predictions of its impact just began to scratch the surface of today's interconnected world. "The wired connection will no longer seem like a strange way of meeting people," predicted a 1995 article in Fortune. [74]

A major driver in turning the world into a global village is the Internet. Since 2000, Internet World Stats reports that there has been a 362.3 percent growth in Internet usage, reaching more than 73 percent of the population in Asia and more than 50 percent of the population

---

73 Robert Rycroft, "Technology-Based Globalization Indicators: The Centrality of Innovation Network Data," *The George Washington University, Elliot School of International Affairs*, October 7, 2002, http:/gstudynet.org/publications/OPS/papers/CSGOP-02-09.pdf.

74 Andrew Kupfer, "Alone together will being wired set us free? Networks will obliterate the industrial model of society. The fear is that they will destroy solitude, and with it human intimacy." Fortune (1995). http://www.cnnmoney.com/..

in Europe.[75] Almost two billion people, more than 25 percent of the world's population, are now connected to the Internet.

Mobile subscribers are also growing rapidly. Forecasts show that by 2015, 244 million people worldwide will conduct financial transactions on their mobile phones.[76] No other media channel offers anything close to this extensive reach.

Among those who have moved to take advantage of this technological interconnectivity are global financial services players such as ING Direct, USAA, Discover Bank, Ally, and HSBC Direct. They have raised the bar with high-yield rates that they primarily market through their user-friendly Web sites.

ING and USAA are examples of the type of new competition financial institutions are facing. Both had remarkable growth attributable almost exclusively to an Internet-based delivery model. At year-end 2000, ING had a total deposit base of $651 million; at the end of the third quarter of 2009, ING deposits were $74.5 billion. This represents a compound annual growth rate of nearly 70 percent. In the same period, USAA grew its deposit base from $6.9 billion to $32.8 billion, a compound annual growth rate of 19 percent. For comparison, all U.S. banks grew at a compound annual rate of 7 percent in this same period. Both ING and USAA have accomplished this growth without the traditional branch infrastructure utilized by most banks. ING offers no traditional branches but does offer seven Internet cafés in the U.S. USAA offers one traditional bank location in San Antonio, Texas as well as one financial center, also in San Antonio. More impressively for USAA, it only began offering banking services to the

---

75 Internet World Stats, *World Internet Usage and Population Statistics*, Internet Usage Statistics, http://www.internetworldstats.com/stats.htm.

76 PYMNTS.com, "Mobile Banking Subscriber Numbers Doubling Every Year," February 18, 2010, http://www.pymnts.com/mobile-banking-subscriber-numbers-doubling-every-year/.

broad U.S. market in 2009. Prior to that, its field of membership was limited to military personnel and related individuals.

Perhaps an even more formidable competitor is the one that combines strong Internet banking capabilities with the traditional delivery infrastructure of a bank. HSBC Bank USA is an example of this type of entity.

HSBC Direct, the online bank for HSBC Bank, reported that savings account balances were up 18 percent to $15.5 billion in 2009 from the previous year. In addition to its sturdy reputation, HSBC offers numerous other advantages that most other online banks don't put forward, specifically, a hefty base of brick and mortar banks (8,000 branches worldwide), access to over 395,000 ATMs throughout the U.S. and a strong international presence. While it is not as visible in the U.S. market, it is the largest non-government owned institution in the world and has not needed any governmental bailout through the recent crisis.[77]

The Internet has also created greater interconnectivity for financial institutions with their clients. For the first time, more bank customers – 25 percent – favor doing their banking online instead of any other transaction modes according to a survey by the American Bankers Association.[78] The Internet is the preferred banking method for account holders under the age of 55. In contrast, the survey reveals the popularity of ATMs has fallen in all age groups. Through 2012, the number of online banking users is projected to develop at a compound annual rate of 20 percent, according to a report from financial services

---

77 Becky Yerak, "Online banking surpasses branches as preferred method for transactions," *Chicago Tribune*, September 24, 2009, http://archives.chicagotribune.com/2009/sep/24/business/chi-tc-biz-onlinebankinghedsep24.

78 American Bankers Association, "ABA Survey: Consumers Prefer Online Banking," September 21, 2009, http://www.aba.com/Press+Room/092109 ConsumerSurveyPBM.htm.

researcher TowerGroup, which defines online banking as anything occurring at a bank's Internet site after a login.[79]

That is good news for financial institutions that have devoted themselves to creating a viable presence on the Web. The bad news is that they now face challenges from competitors that are more global and have also discovered the Web as an additional way to market to their account holders and prospects.

While these tools can be used to extend reach and operating efficiency as well as allow institutions to exploit their differences, they also allow new entrants that can innovate quickly. This is further evidence of the importance of financial institutions reclaiming their historical role as a primary, trusted financial intermediary. To do this they must embrace their regional competitive advantages while reflecting the demands of an increasingly connected and knowledgeable global marketplace.

By understanding what their target account holders' marketplace options are, the financial industry should be in a better position to create a strategy that differentiates itself and exploits its unique community affinity. Recognizing that there are more choices, allows each institution to craft a strategy that is sustainable and will win in the long run.

Cost Considerations

It is costing more to run a financial institution that ever before. It is not going to be feasible for all financial institutions to deal with the credit crisis, stay connected to their account holders, face margin compression and absorb the increased cost of regulatory compliance and security, and still be profitable enough to stay in business.

---

79 Yerak, "Online banking surpasses branches as preferred method for transactions," 2009.

It is already taking a toll. An average of 10 financial institutions failed each month in 2009, nearly four times the number that failed in 2008. The bank failures, 140 banks totaling $170.9 billion, are the most in any year since 181 financial institutions collapsed in 1992 at the end of the savings-and-loan crisis.

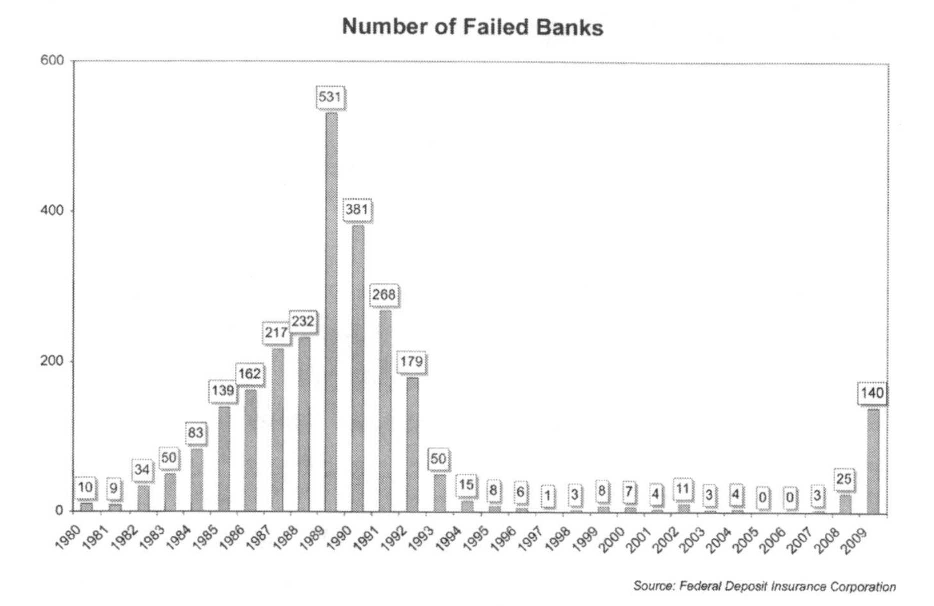

Bank and savings institution failures peaked in the late 1980s. For most of the 2000s failures were an extremely rare event. That changed in 2009 and is likely to continue on an upward trend.

When a bank fails, the Federal Deposit Insurance Corporation swoops in, typically on a Friday afternoon. It tries to sell the bank's assets to buyers and cover its liabilities, primarily customer deposits. Any remaining costs are covered by the FDIC deposit insurance fund, which is designed to be funded by the industry.

The Federal Deposit Insurance Corporation, so sapped by the wave of collapsing banks, fell into the red as well during this crisis. Bank failures cost the FDIC's deposit insurance fund an estimated $25 billion in 2009 and are projected to cost an additional $100 billion through 2013. To replenish the fund, the agency will require financial institutions to pay in advance $45 billion in premiums that would have been due over the next three years.[80]

Smaller financial institutions are in danger of becoming extinct. According to FDIC data, the number of small banks and savings institutions, defined as having less than $100 million in assets, has decreased from 9,455 in 1992 to 2,913 as of September 30, 2009, a decline of nearly 70 percent. Small banks represented 68 percent of all banks in 1992; they now represent only 36 percent of all banks. Similarly, total deposits in these small banks have declined from $358 billion, or 10 percent of the banking industry total, down to $132 billion, only 4 percent of the banking industry total. And whereas this group of banks employed almost 209,000 people in 2009, they now employ fewer than 54,000 people.

Now, the first inclination is to think that this decline in the number of small banks is simply a result of most growing past the $100 million size barrier in this 17-year span. The reality is somewhat different. Of the 9,455 small banks in 1992, only 51 percent still existed in 2009. Half of those had become larger than $100 million; the remaining 49 percent of small banks from 1992 simply no longer exist – most likely having been acquired or merged into another financial institution. The following chart illustrates the decline in the number of small banks from 1992 to 2009. This is contrasted with the number of banks with assets between $100 million and $5 billion, the mid-size "community" bank.

80 Federal Deposit Insurance Corporation, "Banks Tapped to Bolster FDIC Resources," September 29, 2009, http://www.fdic.gov/news/news/press/2009/pr09178.html.

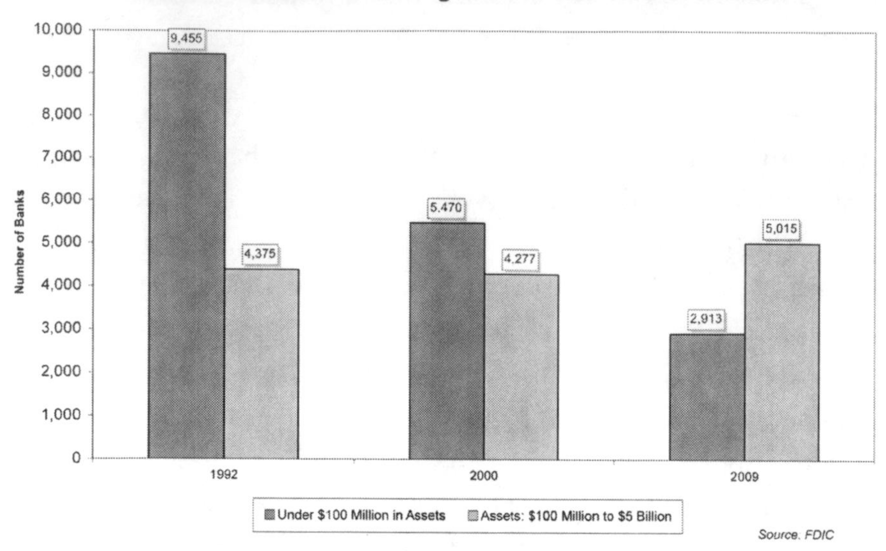

The number of very small banks (under $100 million in assets) has declined dramatically since the early 1990s. However, the universe of mid-size "community" banks ($100 million to $5 billion in assets) has shown growth.

The survival challenge experienced by the very small does not extend up to mid-size community organizations. In fact, the number of banks with assets between $100 million and $5 billion - the mid-size "community bank" - grew from 4,375 in 1992 to 5,015 in 2009. Total deposits in these organizations grew from $1.6 trillion to almost $1.9 trillion, and loans grew from $1.1 trillion to $1.6 trillion. Clearly, this size institution has been able to hold its own even as the overall industry has seen a significant reduction in the number of competitors. The following chart illustrates the growth of loans in the mid-size community bank.

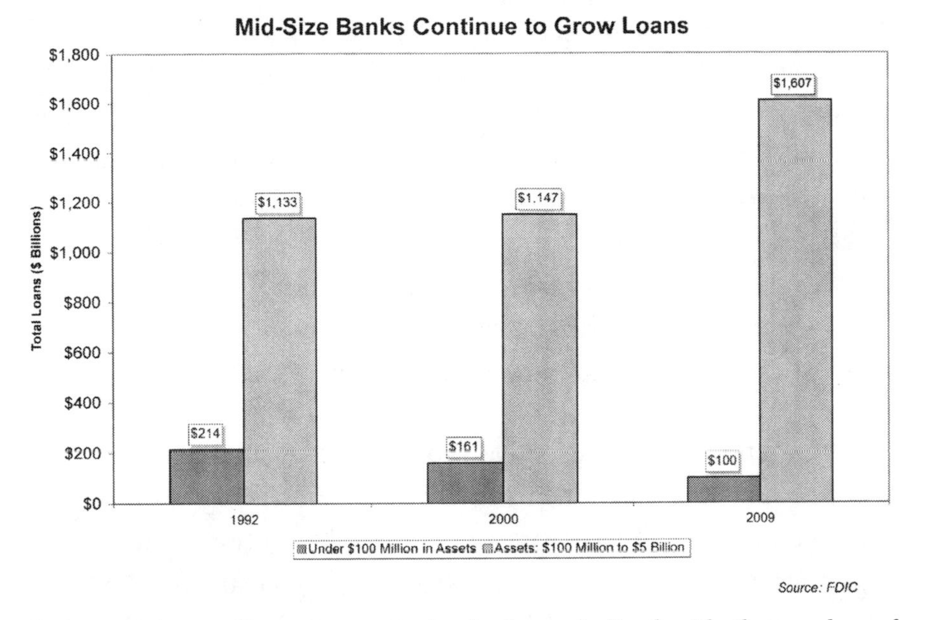

Loans at the smallest community banks have declined with the number of institutions. However, mid-size community banks grew total loans from $1.1 trillion to $1.6 trillion from 1992 to 2009.

Credit unions are not immune to this threat to the very small. From 1994 to September 2009, the number of small credit unions, defined as having assets of under $100 million, declined from 11,606 to 6,408, a 45 percent decrease. Again, this decline in the number of small credit unions is not a function of growth past the $100 million dollar barrier. In fact, of the 11,606 small credit unions in 1994, 54 percent are still in existence and remain under $100 million, and only 7 percent grew past the $100 million mark. The remaining 39 percent have ceased to exist, having been acquired, merged, or closed.

Mid-sized community and large financial institutions are capable of creating a more scalable, operational model due to economies of scale. A Small Business Trends report notes, "Running a bank requires a certain amount of scale, and that floor is rising due to increasing regulatory requirements, channel support, and product support."[81] Consequently, there is a lot of pressure for financial institutions to merge and get larger, so that they can get more efficient and profitable. Put another way, they may not be able to afford to stay small. So the industry needs to face a reality that requires a more dramatic and creative way to address the underlying and increasing cost structure of historical revenue streams under pressure. The answer will likely include a more efficient and flexible operating infrastructure, increased industry collaboration and creative leadership.

The trends noted above, as well as others, have put a severe constraint on spending at a time when the business of banking is becoming more complicated. There was a time when institutions could focus on either increasing revenues or shaving more costs and the hope was to grow into a cost model that was sustainable. Because the basic business model for banking is under so much pressure, leaders of financial institutions are forced to deal with both ends of the income statement at a time when their operating flexibility is lower than it has been historically.

In speeches around the country, I often highlight to boards that it is much different running a financial institution today than as recently as just 10 years ago. The combination of competitive pressures, margin compression, channel proliferation and regulatory, fraud and security costs have left almost no margin of error for today's leaders. An industry that once included stable careers sprinkled with late lunches and golf outings, has given way to a fiercely competitive environment where only the strong and savvy survive.

For community-based institutions, there must be a more innovative culture combined with good risk management so the basic business

81 Small Business Trends, "R.I.P.: The Death of the Small Community Bank," February 3, 2009, http://smallbiztrends.com/2009/02/death-small-community-bank.html.

model can be executed more efficiently and effectively. I don't think this can be accomplished with another round of layoffs. The industry must address its infrastructure issues, deploy technology more aggressively and work collaboratively to lower costs in common processing areas. Much of the back office of a financial institution can be automated and portions can be shared for processing purposes to a much greater degree. Any savings should be spent on developing products and services that are tailored to the account holders' needs in a way that fits with the institution's core strengths and experience. I am, of course, hopeful that by working with congressional leaders, regulators, industry groups and thought leaders, we will see a streamlining of the regulatory framework that balances risk management, safety and soundness, changing market dynamics and is practical about the costs of compliance. However, many leaders don't have the freedom or flexibility to wait and must find a way to set a new course within the existing framework now.

The next few sections of this book will discuss the role and impact of technology on our industry and lay out a plan for all financial institutions to navigate forward.

3: Technology Has Always Had an Important Role

> ***"The first rule of any technology used in a business is that automation applied to an efficient operation will magnify the efficiency. The second is that automation applied to an inefficient operation will magnify the inefficiency." – Bill Gates***

Financial institutions have a rich history of technological breakthroughs and have often set the example for computational innovation. The financial services industry has also been remarkably consistent in continuing to increase its spending on technology and remains the dominant consumer of technology.

Today, financial institutions still spend a significant and increasing amount on technology. In fact, Celent senior analyst Jacob Jegher reports that global information technology spending by financial services institutions will grow to $364.5 billion in 2010.[82] Not surprisingly, this is reflective of increasing demands for regulatory compliance, fraud and security issues, core replacements, emerging payment technologies, mobile applications and business intelligence technologies.

Much of the backbone of service delivery and processing is reliant upon a strong technology infrastructure for financial institutions. From processing high volumes of transactions over multiple channels and product types, to mobile and Internet technology, fraud and security needs, regulatory compliance, predicting service needs and

---

82 Celent, "IT Spending in Financial Services: A Global Perspective," January 9, 2009, http://reports.celent.com/PressReleases/20090109/GlobalITSpending.asp.

cross selling opportunities, lowering costs, staying current with emerging payment trends and social networking interaction, the business of banking is tied closely with technology and innovation. As a result, some of the most advanced technologies continue to be deployed in the financial services industry. In spite of this, financial institutions are also saddled with some of the oldest technology of any industry, particularly with their enterprise applications.

While other less regulated industries evolved to new architectures and platforms in recent decades, the financial industry instead began layering on newer technologies as a way to mask the weaknesses of these older systems while attempting to meet their evolving strategic needs. Regulatory disincentives for change, risk avoidance by the industry, cost of change and pure inertia all combined to create an albatross for the industry. The result is a mountain of layers and layers of applications held together with middleware, proprietary code and luck.

This has left our industry ill-equipped to manage a rapidly changing landscape. And while our industry continues to spend heavily and use some very advanced technology, we are only beginning to address the fundamental issue of legacy technologies hiding under all the newer technologies and costing the industry billions. In this era of smart client, C# (pronounced C Sharp) development language, .NET frameworks, Java, J2EE, Service Oriented Architecture (SOA), Web 2.0, Facebook and Twitter, cloud computing, and human-factors engineering, much of the technological investment in our industry goes to maintaining the old infrastructures and technologies. Too many banking systems manipulate their technology around the same proprietary legacy technology that had its roots in the pre-disco era.

I find that many management teams and boards of directors don't even realize the extent of the bailing wire and tape holding the underlying processing systems together through layers of middleware and other applications, and often only see the new front-end applications. They are unaware of the inefficiency and information risk that lurks until it's too late.

Thirty or 40 years ago, financial institutions were among the most aggressive users of leading-edge technology. Some of the most important processing innovations in the history of computing technology came from the banking industry. Many of the initial enterprise banking systems focused on automating and processing lower level manual transactions as efficiently as possible. As computing technology advanced, this primary demand was largely met. The market needs then shifted towards requiring more than simply automating transactions, but to delivering more information about the person, so the individual could be served better and presented products and services that were relevant to their needs and stage of life. From competition, new product needs, regulatory changes, and the rate of change accelerating in our industry, the demand for greater flexibility increased as the market became more dynamic. This, combined with the need to lower costs and make technology easier to deploy and use, started the demise of legacy systems for most industries.

Underneath it all, the most significant, costly, regulated and highest impact technology platform in the financial institution is the core enterprise application. Almost all transactions through any channel and any product eventually have an impact on the core enterprise application. As the industry begins to see stability in the market, and the bloom of cheap credit and higher fee income ends, many financial institutions are finally ready to more creatively address these issues.

There is an emerging trend to replace these older systems around the globe as boards and executives realize their biggest opportunity to create operating flexibility at a lower cost may come by moving to more contemporary systems. In fact, Baird reports that spending specifically on core processing and related solutions will continue to grow in the low single digits for 2010.[83]

This remarkable spending consistency has become the standard expectation for the industry. Depressingly, most institutions have come to a view that it costs more every year just to maintain the business and keep the lights on, so to speak. The underlying model of layering on newer technologies to mask the older core systems has become an accepted practice, and with it, the expectation that the most expensive and complicated application is not a useful strategic tool to enable strategic execution.

In fact, TowerGroup expected replacement IT spending to increase by over 20 percent in 2009 as institutions consolidated redundant infrastructures and updated mission-critical legacy technology platforms for greater business agility and operational efficiency in managing risk, meeting regulatory standards, and reaching new customer segments.[84] This further demonstrates the frustration and increasing pressure on what had been a nagging issue that has now turned into a severe disadvantage as the competitive and financial pressures have intensified. Many institutions increasingly believe they will be unable to adapt to the changing marketplace quickly or economically with existing systems, in spite of adopting newer technologies as front-end systems.

Many in the industry have begun bundling products across the enterprise to lower vendor costs. Unfortunately, this simply makes them more invested in these older systems, even if there was some immediate cost relief. Replacing these layered legacy systems is the

---

83 David J. Koning and Timothy Wojs, "Business Process Outsourcing: Financial Institution Core Processing Survey," Baird, *December 2009.*

84 TowerGroup, "New Research Finds That Cuts to Valuable Technology Projects Will Place Financial Institutions at Risk," January 28, 2009, http://www.towergroup.com/research/news/news.htm?newsId=5100.

best way to optimize interoperability and customer relationship management across all channels for financial institutions. While languages such as Report Program Generator (RPG), COBOL (Common Business Oriented Language), and Programming Language 1 (PL/1) were revolutionary for their time, cobbling together systems today to operate at top efficiency with newer systems, architectures, programming tools and development approaches is asking programmers to use a band-aid on a problem that needs major surgery.

Many will point to the newer technologies in their enterprise as an example of their commitment to keeping pace. Unfortunately, this intellectual dishonesty does not address the real issue of outdated underlying legacy systems, which must be addressed in order for the industry to make the kind of progress needed for the next phase. In some cases, the layering on of additional applications was the only way institutions could keep pace in order to meet a time sensitive strategic need.

Most institutions have approximately 50 percent of their operating costs in people and roughly 20 percent in systems and support. With the decreased operating flexibility in the business model, many are recognizing that a more dramatic change will have to take place in order to recreate flexibility in the operating model to invest in the future.

Replacing the albatross of these older legacy systems may be the best way to address the biggest cost areas of people and systems while allowing institutions to meet their strategic goals. By doing this, they can then begin to more fully capitalize on several benefits, such as exploiting all the information they have about the people they serve through the same application as compared to separate data warehouses, business intelligence and CRM tools. Those tools help, but are not as powerful as they could be due to their flat file legacy core systems, disjointed systems and duplicate and outdated information source limitations.

Before we discuss how financial institutions can use their enterprise platform as a key strategic enabling tool, we need to take a brief look at how we got here.

A Look Back at Technology in Financial Services

The financial services industry has enjoyed an impressive history of innovative technology. Our industry relies heavily on gathering, processing and providing information, so it's not surprising that banks were among the earliest adopters of automated information processing and were one of the earliest proving grounds for new technologies.

Think of the way banking was conducted before being supported by technology. You would come into the bank office to conduct a transaction. The banker would go to a filing cabinet to pull your account ledger card and carry it back to the front office. They would update that ledger card, manually writing in each deposit, withdrawal, payment of interest; calculate resulting balances (twice to ensure accuracy) then write in the updated balance. They would sign the card. You would sign the card. Finally, the banker would copy the transactions into a daily transaction ledger. (Dual entry, even back then!) Imagine how the introduction of adding machines, or even carbon paper were huge technological advancements.

A Time of Revolution

Computer automation in the U.S. banking market became commonplace in the 60's as bankers realized that many labor-intensive processes could be automated using the first affordable mainframe and mini-computers. These core systems were designed to automate the accounting function and keep track of both account balance changes and transaction history activity.

Because these systems were designed to automate the previously resource intensive functions of account activity management, these new systems were, by design, account centric; an automated replacement for ledger cards. And because banks were operating very separate business silos, each product-supporting system was designed by separate business groups – checking systems for the checking department, mortgage loan systems for the mortgage department and so on. The result was many separate applications with data residing in what is referred to as flat-files that were supported by mainframes and mini-computers of the time. Basically, this resulted in several separate file-driven systems.

In 2008, there was an average of more than 39 million commercial items processed by the Automated Clearinghouse, resulting in more than $62 billion average daily volume.[85] The ability to use computer processing to handle transaction processing was a huge and necessary advancement for financial services and very rightly was one of the first uses for the processing power that these systems could support. Without it, financial services would simply not be what they are today. Now fast-forward to the first wide spread transactional systems.

---

85 Board of Governors of the Federal Reserve System, *Commercial Automated Clearinghouse Transactions Processed by the Federal Reserve—Annual Data*, Automated Clearinghouse Services, http://www.federalreserve.gov/paymentsystems/fedach_yearlycomm.htm.

Flat Files

```
110887600055067MAXWEL RANDOLF SMITH 455 NORTH STREET EAST CAMBRIDGECONN0525037
110887780008675LAWRENCE BROWN        12 MAIN STREET   WINDSOR CT            0609508
110888300048943LAWRENCE BROWN        12 MAIN ST       WINDSOR CONN          0609537
110889400548734SUZANNE DOYON         86 ELM STREET    NORTH HAVEN CONN      0647337
110894575338741LIZ LUSSIER           5 PLEASANT VLLY  SO WINDSOR CT         0607408
```

The flat file data structure pictured above was typically proprietary to the hardware and operating systems. Data files contained one row of data for each account and the programs accessing the data were dependent on understanding each "space" of the fixed length records. Programs would count from the beginning of each record in order to obtain the data elements needed. Codes were used (instead of natural words), in part, because of their ability to take substantially less space within the fixed length records. For example, a three-digit code could represent 999 different meanings or instructions.

Programming

Programming languages such as COBOL or PL/1 were used to create these applications. The programs were very time consuming to write, with commands originally being imprinted, stored and read from card stock. In order to read and write to the data elements within the flat files, the programs needed to know exactly where in the file record the data was located. (In the flat file example, the first 8 data elements are the account number, followed by the balance.) With such systems, if the account numbers exceeded the 8 digits allowed and needed to be changed to 9 or 10 digits, then all other programs had to be updated to reflect all the other data's new positions in the longer records. As client wealth grew, the length of balance and transaction amount fields had to increase. Each time one of these changes occurred, it required extensive programming changes to the core enterprise and reporting systems.

User Interfaces

User interfaces were certainly not those we are accustomed to today. CRT (cathode ray tube) terminals displayed data in much the same way as the cryptic data files. Because there were separate systems for each type of product, employees would need to sign on to separate applications in order to check balances or edit records of different types of accounts. Because the applications used codes (instead of natural language), employees using the applications had to keep track of dozens of different types of codes to complete their processes (20 might mean deposit, 60 might mean withdrawal, etc.).

Regardless of the complexities, these advanced applications were a tremendous success. Transactions were processed faster with less manual effort and higher rates of accuracy than ever before. As the hardware became more affordable, even community banks and credit unions began adopting the technology in-house, instead of outsourcing or utilizing the services of larger correspondent banks to conduct this processing on their behalf.

For almost 20 years, these systems supported the growth and success of financial institutions. In the early 80's however, these systems began to show their many limitations. The period between 1980 and 1994 saw more legislative and regulatory changes in the financial services industry than any other since the 1930s.

Financial institutions began expanding their product offerings. Savings banks, community commercial banks and credit unions all began to offer similar product lines to consumers, all hoping to gain more wallet share with their consumer clientele. Trying to make consolidated banking easier for consumers was a strategic focus for many organizations.

In fact, many vendors still offer several systems focused on specific financial institution types or charters, such as savings banks, commercial banks or credit unions. These initial design limitations still apply. Banks, especially, began to try to minimize the impact of their internal, vertical, product support structure to their clients. But it was difficult to hide that a bank had very different customer information collected and used by its separate business lines.

The Addition of the Central Information File

Because of these separate, line-of-business focused systems, account statements were mailed for each account separately, and many times to different variations of a name and mailing address. When making changes, account holders often needed to work with the organization several times to ensure that their checking account, savings account and mortgage were all titled and addressed the same way. Joint account processing held even more complexities as many of the product applications had severe limitations with the number and types of relationships a consumer could have on an account. Many times these extended relationships were known on the backup paper account agreements only.

Consumers grew tired of having to work at maintaining separate relationships with one organization for their mortgage loans, car loans, certificates of deposit and checking accounts. Even for the bank clerk it was nearly impossible to ensure that all accounts were updated properly. They needed to search each product system separately for the existence of an owned account, update each account separately and conform to very high levels of data input accuracy between accounts.

This fragmentation of the consumer's data was recognized as a flaw of the early product design. Banks needed a single source for customer information in order to use these primary records and to manually (later automatically) update this information on the individual account records. An additional file system to unify person and business information was added – the Central Information File (CIF) was born. These new file systems held unique CIF keys to identify customer records. Keys were usually a cryptic combination of letters and numbers used to link an individual's information across multiple applications files. Credit unions undertook similar strategies with membership records containing a member number. In both cases, the approach was to create records to maintain the primary source of name, address, tax identification and account relationships.

The CIF created new possibilities for significant improvements, both operationally and for unifying the account holder's experience. For the first time, front-end systems (which will be discussed later) could display summary information of all accounts at the person level. Combining accounts on single periodic statements saved paper, printing, postage and labor costs, while improving satisfaction levels with the financial institution. The CIF was the first technology step taken to unify the business silos within the organization and provide a single interface to the account holder.

Unfortunately, with the introduction of the CIF, massive clean-up strategies were also spawned. Like before, each account might have separate name and address information for a given account holder. The introduction of the CIF was an additional version of this information. The CIF could not always delineate if the person whose record was held in it was the same person from one by the same name in the separate account systems. User coordination, decision making and manual linking between the CIF and account records was necessary. Unfortunately, users were not always accurate or thorough in their determinations. Reviewing, cleaning and linking the data between accounts and CIF systems is an annual (or more frequent) process that many organizations still engage in today.

Front-end Applications

The introduction of personal computers also had a major impact on this process. With them, the first DOS, Windows and OS/2 based bolt-on teller and platform applications were launched, allowing a better user experience compared to "CRT green screens" that were native to the system.

Companies specializing in teller systems and new account systems or platform systems began to emerge to capitalize on this new market. The new systems were designed to be easier to use for front-office staff rather than the green screens preferred by operational staff. These new applications translated codes and abbreviations into pictures and more recognizable words designed to make front-line staff more efficient.

In order for these new graphical user interface (GUI) screen platforms to communicate with the older technology of the core systems, data translating middleware was required. This added yet another layer of complexity by requiring manual updates or batch translating programs to replicate the core data in the teller and new account system. Transaction codes, products codes, rates codes and service charge codes needed to be established in the front-end systems so that the translation could occur.

Many times vendor updates to the core system caused operational issues in the front-end systems when new releases were distributed. Not only was synchronization required between the data in both systems, but also between release schedules with multiple vendors since the teller and platform systems were designed to work with many different core systems.

Soon there was an even greater shift to customer centricity, further exposing the gaps in the original product based design. New, more flexible systems to manage customer knowledge were needed.

The introduction of Customer Relationship Management systems, or CRM, added an additional information layer that was more adaptable to the changing informational needs between financial institutions and their account holders. All of these third-party front-end applications, including teller, platform, call center, and CRM systems, allowed financial institutions to overcome the usability gaps by translating core transaction codes, product codes and other cryptic information for easier use by front-line employees. However, they further exacerbated the flaws of the underlying legacy systems that were designed around account transactions and not people. These multiple systems layered on top of the legacy core added benefits to tellers and platform staff, but they created the need for more quality management efforts to ensure that client data across all applications was accurate.

While the original legacy systems and front-end applications that followed helped to automate the functions of account management, we still find that what many financial organizations are relying on today is a layering of flat-file systems and third-party application wrappers created to extend the life of the aging core system. Imagine if another industry was working on the same antiquated technology as banking systems. What if the entertainment industry was still trying to use record players to revolutionize the way we listen and enjoy music?

Customer information resides in a complex combination of product files, CIF files and third-party CRM applications. Maintenance of transactions, products and services must occur on each system. And does it truly support the business model of consumer centricity that the industry is trying to achieve in today's marketplace?

The Business Model Changes

Late in the last century, financial services saw many new entrants that fundamentally changed the face and attitudes toward banking. Internet banks, insurance companies and even retail giants such as Walmart redefined the competitive marketplace. In order to compete with these new client experience-focused experts, banks and credit unions needed to sell relationships and convenience as much as, if not more than, simply financial accounts. For the first time, the business of banking was being measured against service giants like Nordstrom's and L.L.Bean.

Unfortunately, the systems supporting the more common brick and mortar organizations were not designed to focus on relationships and the quickly changing channel strategies needed to support consumer and commercial convenience strategies.

Changing data structures in the flat file systems created unmanageable challenges for legacy providers supporting applications with dated programming languages, and caused severe limitations to trained workforces.

The layering approach taken by legacy core providers did not support the flexibility, data clarity, and time to market that today's competitive financial institutions demand. Most legacy vendors admitted they would not write a new system because it would cost too much, take too long and be too disruptive.

The Introduction of New Banking Technologies

In the 80's and 90's, newer technologies and data storage solutions became commonplace. Proprietary databases, operating systems and antiquated coding languages were no longer used when creating enterprise applications. Many industries took advantage of these new database structures, including the Relational Database Management Systems (RDBMS).

Unlike a database program that stores and retrieves data, an RDBMS shows the structure and the organization of data within the database. It allows a more flexible and expandable means of storing data using "many to many" types of relationships.

Legacy Flat File Method

As a simple example: A person might have many different addresses. They might use a PO Box for mailing, one residence address; have many vacation addresses and even a work address. In legacy, flat-file storage methods the number and types of addresses collected must be pre-determined. So, those systems might choose to hold two or three addresses.

Person Record

```
MAHERN,Mike Ahern,786776118,11171958>
AAHERN,Alice Ahern,7867854417,06311962>
```

Address Record

```
MAHERN,PO Box 180,Simsbury,755 Main Street,Simsbury, 985
Beach Road,Oak Bluffs>
AAHEERN,755 Main Street,Simsbury,755 Main
street,Simsbury,,,>
```

RDBMS Method

In a relational model, the address table supports unlimited addresses. Person and address information are contained in separate tables. A third table ties the information together. So one person may have unlimited addresses and one address may be related to unlimited people. This is the "many to many" relationships that allow these types of databases to be self expandable.

Person Table

Person	First Name	Last Name	Tax ID	Date of Birth
1	Mike	Ahern	786776118	11/17/1958
2	Alice	Ahern	786785441	05/31/1962
3	Mark	Jones	557315844	07/28/1992

Address Table

Address	Line 1	Line 2	City	State	Postal Code
85	PO Box 180		Simsbury	CT	06026
86	755 Main Street		Simsbury	CT	06026
87	985 Beach Road	Unit 3	Oak Bluffs	MA	07035
88	5865 36th Street	Apt. 475	New York	NY	04028

Person/Address Relationship Table

Person	Address	Address Type
1	85	Mailing
1	86	Primary Residence
1	86	Vacation 1
2	86	Primary Residence
1	88	Work

Many industries deploy relational databases because they are cost effective and efficient due to the simplicity in programming, access and maintenance. Educators use relational databases for archiving and storing student records. Accounting applications like QuickBooks use a relational database engine for linking accounts, vendors and transactions. Airline and hotel chains, hospitals, retail and Internet search engines are all industries that rely heavily on relational databases.

Today, data security and integrity are so crucial that they are among the most regulated areas in the financial services industry. Products such as IBM's DB2, Microsoft's SQL and Oracle's RDBMS are clear global leaders providing database solutions that support advanced data management requirements. For instance, Oracle offers real-time data redundancy that is "best of breed" for disaster recovery and other real-time data redundancy needs. Encryption is offered at the database level, fully securing stored data. Tools are provided to improve the data manipulation as well as the speed of development.

Applications using these advanced database technologies are faster, more secure and more reliable. Some database products, like Oracle, can run on many different hardware options and operating systems, providing the ultimate flexibility in deployment. Because these database solutions focus on data management tools, application providers can concentrate on creating the data models and applications that best serve the industry.

Unfortunately, it is not enough to simply have a relational system. It has to exploit the full advantage of relational, real-time, non-redundant data with the most efficient application tools without all the layers masking the underlying legacy system. For instance, a vendor may port a legacy system to run on a relational database but not modify or exploit any of the underlying benefits described above. This example would serve only to market the upgrade. With no investment in exploiting these advanced tools through a commitment to an advanced data model, there would be few benefits. Another example might be a system that was designed with an account structure geared to a credit union, thereby limiting how far it could be used for commercial banks or internationally without more proprietary code.

Why Should It Matter to Today's Financial Institution?

In stark contrast to the expense and risk of staying the course with an aging legacy system, core processing built upon an open and flexible relational architecture at the core level enables financial institutions to meet their strategic goals without the addition of multiple layers of technology. These are proven, viable technologies with a solid record of accomplishment in terms of installations and performance. The advantages of more modern core solutions are now too compelling to ignore:

- Scalable to any size financial institution
- Field size flexibility for all numeric and text field types
- No limitations to the number of products or the number of fields
- Manages flexible relationships between data elements: people, business, accounts
- Hardware independence
- Operating system choices: Windows, UNIX or Linux
- Industry standard development and integration protocols: OFX/IFX, XML and .NET
- Service Oriented Architecture
- Delivery independence: SaaS model, data center, or in-house

An enabling enterprise platform begins with an open relational database architecture designed to mirror the business processes of financial institutions. Server level or PC applications interrogate the database directly to acquire all data from the single data source. And finally, an application layer drives the business rule functionality of the financial institution.

Y2K brought an unexpected value as it sharpened the focus on the advantages of a more advanced design and an open file/field structure at the database level. Industry regulators were quick to recognize the integrity and advantages of relational real-time systems. Likewise, Y2K shined a bright light on the hazards and complexities of trying to maintain archaic programming languages like COBOL or PL/1 and layered software systems.

Rather than try to survive the weaknesses of legacy platforms and the ineffectiveness and cost of additional layers, financial institutions recognize there are now better options available at a lower cost. Many forward-thinking financial institutions have taken the step to open, next-generation technologies to replace old legacy systems. These institutions recognize that the total cost of ownership for the newer technology far outweighs the risks involved in maintaining their aging system that limits growth, impairs productivity, minimizes new opportunities and puts the institution at risk.

Legacy mainframe computing has often been synonymous with scalability. But now, Windows, UNIX and Linux platforms offer comparable performance benchmarks at a much lower cost with an added benefit of superior ease-of-use for the end-user, as well as a growing pool of information technology staff.

Why should you care about the architecture and technology? What does all this scrambling mean to you?

Financial institutions should benefit from their core enterprise system. The core system should be person-centric, have no data redundancy, and be open and flexible. Today's systems do not need codes. Information should be stored in real words so there is unparalleled consistency between the database, reports, statements, notices and front-line screens. A system should not have limitations to the number of products you can offer, or the number of different types of service charge routines or interest rate parameters your financial institution can set up. The system should be capable of tracking unlimited, intricate consumer relationships and never require CIF data scrubs. Extensibility allows even greater flexibility in meeting the changing needs of an institution's client base and competitive landscape. The architecture should also allow for continuous upgrades to the latest application tools so it never gets outdated.

The Transition Has Begun

Financial institutions relying on legacy core systems, some of which date back to the 1960s and 1970s, have gradually become more aware of the limitations of their archaic legacy technologies with bolted-on applications that camouflage their intrinsic limitations.

This chronic issue has become a severe competitive disadvantage for financial institutions seeking to implement client-focused services, diversify revenue streams, increase fee income, and introduce new products and services more quickly, and at a lower cost.

Financial institutions are experiencing unrelenting demands to know more about their clientele, improve service, increase wallet share, create stickier relationships, and keep up with regulatory changes, all while trying to continually improve operating efficiencies. Many North American financial institutions still function within the old legacy environment, and find they have reached a threshold in terms of how much more efficiency they can extract from their core data processing systems. They often don't realize that there is a different model or that there are significantly different choices available to them. They've become so accustomed to working around archaic limitations, such as data field sizes, limits on a fixed number of product types, redundant databases, antiquated coding principals and the resulting layered-on system approach to working around these issues, that they don't know how to approach the basic problems differently. They often cannot effectively communicate the serious handicaps they encounter.

Around the world, large financial institutions are on the leading edge of core upgrades, but in North America it is the community-based institutions making the move. And the drumbeat of change is growing louder. A shift toward more open and flexible next-generation technologies at the core level has been ongoing globally as the business demands have intensified. Along with mid-sized institutions in North America and around the world, some emerging nations are leapfrogging the U.S., much like the move directly to wireless telecommunication skipped the hard-wired generation. What's left is a more agile, flexible, and low cost infrastructure.

Large institutions in the U.S. have the budgets to work around the trap of these legacy systems but find it too daunting to employ a complete overhaul. Conversely, community-based institutions can handle a single conversion to a more contemporary system, and have the cultural drive to compete and win. This puts some community-based institutions in the unique position of having a technological advantage with a fraction of the scale or cost.

For example, Union Bank, which Bank of Tokyo-Mitsubishi bought in November 2008, is the biggest U.S. bank to announce a core project.[86]

"By updating their cores, they become more agile. Being competitive today requires more than the typical transaction processing that [legacy] core systems perform," shouted a June 2008 article, *The Heart of it All*, in Bank Systems & Technology.[87]

According to IBS Publishing's *The Retail Banking Systems Market Report*, "Old core systems continue to become less and less sustainable, for both cost and business reasons, but the counterweight in many

86 Michael Sisk, "The Core of the Matter," *American Banker*, October 29, 2009, http://www.americanbanker.com/specialreports/174_21/the_core_of_the_matter-1003410-1.html.

87 Maria Bruno-Britz, "The Heart of It All," *Bank Systems & Technology*, June 2008.

organizations is the market-driven cost constraints that impact major decisions such as core system replacements."[88]

In the face of the economic crisis, it's easy for financial institutions to lose sight of their ultimate goal and begin to feel that cost cutting alone will address these issues. But taking a proactive role in an institution's infrastructure is what will ultimately drive growth. Credit Union Journal reports, "Many credit unions are turning to operational efficiencies to reduce costs, while actually increasing both service levels and their competitive edge. To accomplish this, they're first making an investment – in an open, flexible core system." Credit Union Journal further reports that institution-wide integration is also an efficiency drive that saves time, reduces errors and slims down operating costs.[89]

Financial institutions are often wary of implementing a new core platform – their most expensive, most regulated, and most mission-critical technology – until they observe other successful conversions and can calculate the potential return on investment for their organization.

In addition, many financial institutions are finding that doing nothing brings its own set of risks, particularly in terms of the fundamental integrity of an institution's enterprise infrastructure and its ability to respond to new competitive challenges. Now, more than ever, client retention and profitability, back-office automation, and the ability to build additional value around each financial relationship will be the critical factors for success.

This was never more evident than in the recent financial crisis where the disjointed approach to risk decisions, inconsistent information

88 IBS Publishing, "The Retail Banking Systems Market Report," Edition 6 – 2009.

89 David McCooney, "Why Credit Unions Are Discovering That You Have to Spend Money to Save Money," *Credit Union Journal*, July 27, 2009.

about the same person and the inability to respond to regulatory changes showed the industry it was way over its head.

On the technology front, we are witnessing a global shift away from legacy systems to more open, relational, and real-time technologies. This movement is most apparent in Europe, where the adoption of the Euro common currency has brought a significant impetus for change. New spending for core systems and the implementation of more real time people-centric solutions are occurring faster now. The rest of the world is following suit. Both in the United States and globally, a focus on core replacement is evident.

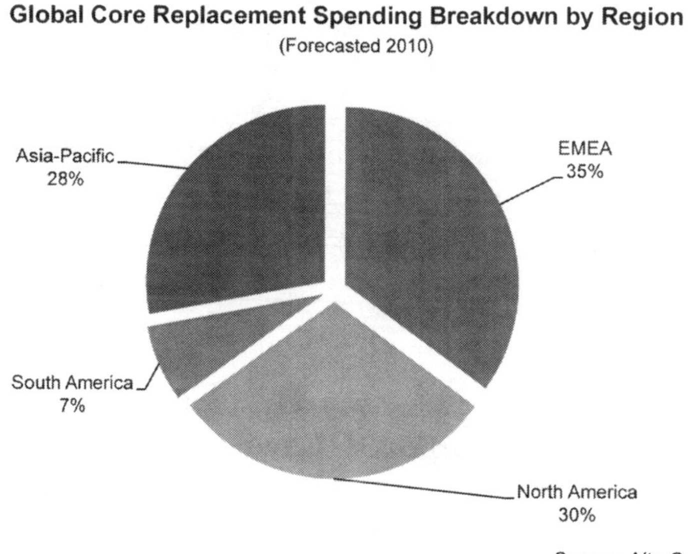

An increased need, as well as a focus on core system replacement, is not just a U.S. trend, but rather a global one.

The graph above demonstrates that global core system replacement is almost equally prevalent in three of the four geographical regions. In a 2008 Microsoft white paper, research finds that, "... In Europe, legacy core system replacement is now firmly on the agenda of a significant proportion of larger banks."[90] Industry analyst Forrester comments "... European banks and the international subsidiaries of non-European banks drove a high amount of new and extending banking platform business in Europe."[91] "Overall, the European retail banking core systems market is set to grow at a CAGR of 6.5 percent from 2006 to 2010..."[92]

With regard to the U.S., research from IDC Financial Insights shows that almost 20 percent of financial institution survey respondents plan to invest in core banking in the next 12 months.[93]

According to Celent Communications' report in *Asian Banking Technology*, interest in replacing core banking systems has been slowly simmering for some time in banking markets around the world. Most banks are running back-end systems based on mainframe technology introduced 20 or even 30 years ago. While these core systems are quite stable, they are not very flexible, making it time-consuming and costly to modify them to handle new products, deliver new system functions, or support person-centric service and marketing strategies. In order to deliver these features, increasingly necessary in the competitive environment of the 21st century, IT departments have been augmenting their legacy processing systems with middleware extensions, auxiliary systems and armies of back-office staff to manually handle, for example, relationship pricing, loan pricing and other detailed processes. Against this background, the Asia Pacific region is emerging as a leader in core systems replacement. Both first and second tier banks in a number of markets in the region are

---

90 Microsoft, "Core Banking with Microsoft Technology White Paper," February 2008.

91 Jost Hopperman, "Global Banking Platform Deals 2008: Regions," *Forrester*, (July 2009).

92 Datamonitor, "Core banking strategies in European retail banking (Databook)," Research and analysis highlights, December 12, 2007.

93 IDC Financial Insights, "Business Strategy: IT Executives Survey Results – Can 2010 Get Here Soon Enough?", November 2009.

installing new core processing systems, often using open platform or distributed technology.

Taken together, this powerful convergence of the global shift away from legacy technology and the intensifying competitive landscape creates both an imperative to change and an opportunity to prosper.

Besides having to spend too much time, effort and resources to support and maintain existing legacy systems, financial institutions are beginning to recognize the extra scope of risk exposure associated with older technology.

The evidence is clear. Financial institutions around the globe are discovering that proprietary software and underlying databases are becoming serious inhibitors to their business-growth strategies. Instead of utilizing technology to open the door for progress, financial institutions combat the limitations of legacy systems that are: not extendable, supported by antiquated coding standards, constricted to inflexible hardware and operating system deployments, and require redundancy of data and business logic between the original applications and the many added layers. At a time when expanding reach and improving operating efficiencies are key initiatives for the financial services industry, these systems simply act as the opposite of "enabling technology."

The Need for Flexibility

In today's quickly changing financial services marketplace, it is difficult to fully define all the current, complex requirements. The opportunities and challenges that will present themselves tomorrow, next year or in the next decade are almost impossible to predict. Yet, technology and software investments can range from 3-to-15-year investments. Ensuring that technology investments will support future business requirements must be top of mind.

What services will you offer in the future? What channels will clients demand that you support? What types of relationships will you want to maintain? What new lines of business will you enter? Who will be your competitors? What new risk factors will you face?

Hardware, operating systems, data maintenance solutions and software together must support today's and tomorrow's requirements. When you consider these forward thinking possibilities, it's important to understand to what extent you can parameterize your current environments. It's important to know the level of difficulty your software providers will encounter to keep up with the pace of change, what level of commitment to innovation that they have shown, and what resources they rely on. It's important to use hardware, operating systems and data management solutions that also provide flexibility, are widely utilized and accepted, and enjoy regular and active investments. Flexibility is a business need today that mitigates risk for tomorrow.

Knowing More about Your Account Holders

Ten years ago, all you needed to know about your client was their name, tax identification number and mailing address to send statements. Regulatory best practices and the need to grow broad based relationships with clients have changed that dramatically. Today, you want to know the residence of your clients as well as their vacation address, email address, cell phone number, family relationships, business relationships, household income, brokerage accounts, Internet preferences and much more. You want to capture any information possible about the person, limited only by what they are willing to share. The purpose of obtaining this information is to serve the individual better, as compared to being limited by the ability of your vendor's product to obtain and utilize this information.

Business relationships require even more complex information. This requires a single system of record, supporting not only the regulatory needs of expanded client data, but also your business objective needs. Tracking intricate relationships between people, businesses, organizations and accounts within your organization is paramount. In no other way can you identify the true benefit of a client's relationship without understanding the full reach of their financial wallet and external relationship influencers.

Openness and Security

What worked in the past no longer supports today's change of pace and risk inherent environments. Organizations change over time. Their technology infrastructure must be flexible to support these changes. Applications should be supported via optional hardware, operating systems and deployment options. Data should be secure at the database level, and real time onsite and offsite disaster recovery of data should be available. Operating systems should be widely supported and offer frequent security patches and application development tools. Vendors should be chosen with the knowledge that their applications are developed and supported by a widely available resource pool. Openness should allow a financial institution the choice to use any hardware, any operating system, and any combination of products anywhere in the world.

Extensibility

If the application and data model are not designed with extensibility in mind, every new data need, new product need, and new feature need will be an expensive and lengthy process. In this case, the data and application design is as important as the technology it is built on. The designers should predict the places where extensibility will be needed in the future.

In this way, an organization is never constrained by limits. They can create as many products, as many pricing routines, as many data fields, as many types of relationships as they may ever need. The same is true for the vendor. Initial extensibility in design, benefits the speed of easily adding new services, new access methods, and new business logic by simply extending the application.

This is particularly important for our industry where increasing cost pressures combined with a culture of collaboration could produce a unique opportunity to navigate the next phase of growth. In order to do this, more vendors will have to allow simple, extensible contemporary tools that do not require significant development resources. Instead, they must offer easy-to-use tools to launch products, multiple languages, etc., as well as easily create applications that can benefit the industry and can be shared around the world at a low cost.

Like the birth of banking, this would allow the industry to recapture the important social, community and economic role that was part of its foundation. This rebirth can occur with the help of leading-edge technology tools that could facilitate the transition to better, faster, less expensive and highly innovative platforms. This model would cause an explosion in innovation as everyone benefits from all innovation investments if they are on an extensible single platform. Applications would beget other applications, all aimed at meeting market needs faster and at a lower price. Institutions would be able to share and collaborate on new ideas and tailor them to their own needs to compete more effectively. This would be much like the applications that have become popular with the iPhone store. And they would be created by peers that are knowledgeable about the industry and could share for free or a fee, the innovation they've invested in.

Now that we've studied how and why the omnipresent factors of technology, competition, regulation and consumer demand are gathering force and playing off one another, here are some valuable questions that can help further define your own financial institution's position and develop a set of navigational waypoints for your institution's future technological direction:

1. Can I get the information I need, when I need it?
2. Does the application serve all account holder touch points consistently through a single application?
3. Am I trying to overcome the problem of having an application that is inflexible by adding another "bolted-on" solution?
4. Am I forced to layer middleware onto my core system in order to gain flexibility?
5. Are there any limitations on file sizes, numeric values or account types?
6. Are there separate "bolt-on" teller platform systems required?
7. Is the solution relational at the core application level?

To be truly open and extensible, the application must be usable throughout the enterprise, not just at specific terminals, to produce and distribute actionable information. The functionality must integrate the workflow and transaction systems without heavy, proprietary interface efforts.

A good way to answer some of these questions is to ask for and then evaluate a high-level diagram of your complete technology infrastructure that highlights each component of your technology infrastructure. Then map each one to a specific targeted business strategy.

At the same time, you will want to assess whether your technology platform is:

- ❑ ***Open***, to allow many hardware choices and easier integration with third-party products or services.
- ❑ ***Relational and Real-Time***, to provide more relevant information immediately to better serve and sell clients the products they seek and need. This would allow you to better meet regulatory reporting changes easily and efficiently.
- ❑ ***Scalable***, to give your financial institution the freedom to change direction and add products and services at a low cost.
- ❑ ***Efficient***, to help boost productivity across all areas of the institution. This also assists in meeting strategic goals without multiple layers of applications that add cost and complexity.
- ❑ ***Normalized***, to provide processing speed, system efficiency, non-redundancy and reliability of data.
- ❑ ***Extensible,*** to have the ability to add information, products and applications in an unlimited way.

The concept of extensibility amid this community of innovation is a key opportunity for financial institutions that have already latched onto the concept of consumerization, as it relates to their account holders.

By having a truly extensible system, all constituents should be able to further differentiate themselves by having a way to extend user applications beyond the traditional means. This would mark an important evolutionary step in vendor relationships and signal acceptance in the more collaborative and interconnected world that we live in.

As an example, in a traditional relationship, the vendor listens to the market and attempts a solution, which is sold and maintained as it improves. The hope is that the product is close to the market need, and that maintenance payments will fund improvements to the system. In a more extensible environment, the vendor provides an easy-to-use tool that allows the vendor, partners and/or clients to work collaboratively to create products and services, and share the developmental tools that extend the existing applications. The network or community works together to create organic solutions in real time. This is an important step in the evolution of application development. When embraced more fully, it will differentiate companies and their clients by driving innovation at a dramatically lower cost.

While the core enterprise system improvement could have the most dramatic impact on today's market as well as individual institutions, there are several other areas of technology that will be important over the coming years. With a properly aligned core application and strategic outlook, many of these related technologies become more powerful and economical to deploy.

IDC Financial Insights expects the top technology spending priorities for 2010 to be focused on improving profit margins, increasing sales, retaining existing customers, and reducing operating costs.[94] I hear these themes echoed around the world, along with regulatory, compliance, and capital preservation.

As a result, we expect a continued focus on the following areas:

Regulatory and compliance spending – With the dramatic increase in regulatory changes and complexity, this area will continue to grow. There are no single, high-quality solutions for community-based financial institutions, as most monitoring and reporting tools are

94 IDC Financial Insights, "Business Strategy: IT Executives Survey Results," 2009.

designed for larger institutions. This bucket of spending will also include the modifications of existing systems and enhancements to meet the changing requirements. The increased focus on enterprise risk management often falls in this area. According to AMR Research Inc., data, total governance, risk and compliance spending is expected to grow to $29.8 billion in 2010.[95]

Security and Fraud prevention and detection – As unquestionably the most targeted industry for fraud and the increased channels by which fraud can be perpetrated, this will continue to be an increased spending area for financial institutions. Along with the cost of adding more channels, comes the investments needed for prevention and detection for those channels. In a 2008 study by Gartner research, results showed that spending on fraud detection was on the rise. About 60 percent of banks expected to spend more on fraud detection and customer authentication in 2009 than in 2008.[96]

Mobile banking – Mobile banking is still in its infancy in terms of what protocols will win and how users will interact with their balances through these devices. However, the significance of mobile banking's growth in popularity and potential is becoming clear. "Mobile banking is growing far faster than Internet banking and is attractive enough to deliver significant numbers of new customers," said Richard Crone, CEO and founder of Crone Consulting, LLC.[97] The significant growth in mobile commerce, attractiveness of the demographic, growth and adoption rate compared to Internet banking, and growth in total subscribers make this a highly sought-

95 Linda Tucci, "Governance risk and compliance spending not focused on technology," *SearchCompliance.com*, December 1, 2009, http://searchcompliance.techtarget.com/news/article/0,289142,sid195_gci1375707,00.html.

96 Gartner, "Gartner Says U.S. Bank Spending on Fraud and Authentication Is Rising," July 14, 2008, http://www.gartner.com/it/page.jsp?id=721009.

97 Richard Crone, *Bankers As Buyers 2010,* ed. William Mills Agency (2010), 17.

after application even though the implications and benefits to the banking model are not completely understood.

In January 2009, ABI research originally forecasted mobile sales of physical goods in North America would reach $544 million, up 57 percent from $346 million in 2008. Not bad for a still fledgling industry. Ten months later, ABI research revised its estimates to $800 million based on mobile commerce activity and predicted it will double in 2010.[98]

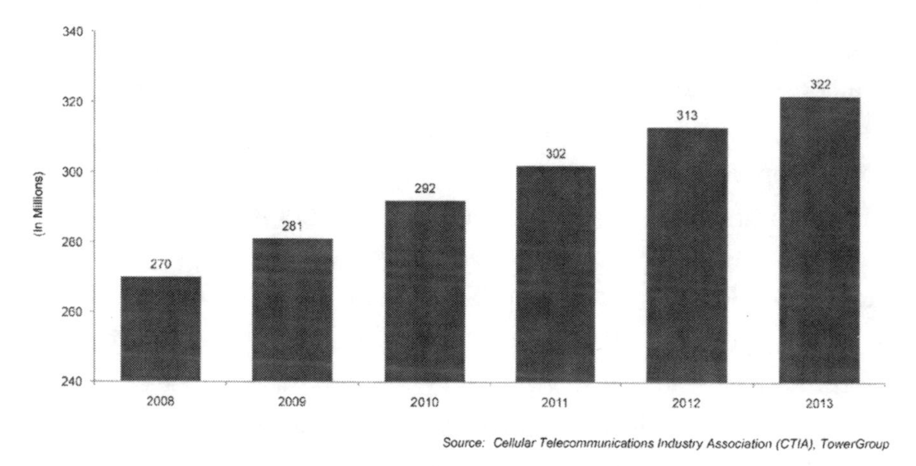

TowerGroup shows that mobile device subscribers have been on the increase since 2008. The number of subscribers in 2013 is projected to grow by almost 20 percent from 2008 levels.

Integration technologies - Between trying to get more out of their existing technologies, making their current platforms more efficient and streamlined, and mergers and combinations, financial institutions will place increased emphasis on technologies that will make their current environment more effective and efficient through tighter integration.

98 Olga Kharif, "M-Commerce's Big Moment," *BusinessWeek*, October 11, 2009, http://www.businessweek.com/technology/content/oct2009/tc20091011_278825.htm.

Business intelligence – While there is a strong focus on cost containment and capital preservation, most institutions that want to thrive more than survive realize that better information about the consumer allows better risk assessment, more targeted pricing and packaging of products, better service through prediction of the needs of the individual, and might enable the institution to capture a higher percentage of wallet share. While short-term needs may be focused on other areas, those institutions focused on making their competitive differences more obvious realize that this remains one of the key areas to unlocking value around service and pricing.

A 2009 Forrester report comments, "Using assumptions from real companies facing the upgrade decision, we found that a business-driven CRM application upgrade can generate a risk-adjusted 31 percent return on investment over five years." [99] Aite Group also predicts an increase in CRM spending through 2012.

Financial Services CRM Apps Spend (US$ millions)

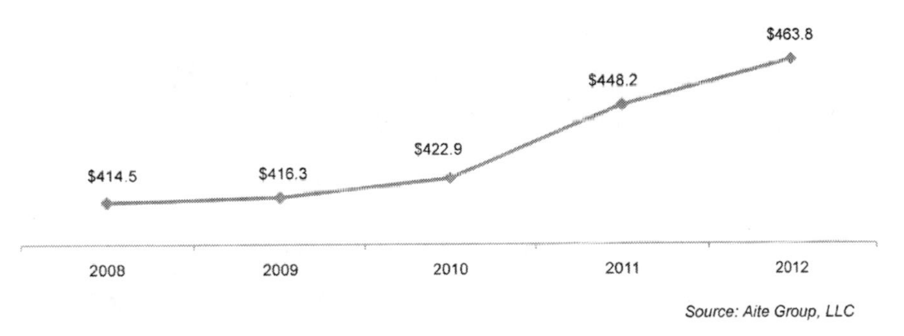

Source: Aite Group, LLC

Aite Group is expecting a continuing trend in CRM spending that will reach over $460 million in 2012. Note the expected dramatic increase in spending in 2012 and 2013. This reflects the continued emphasis on a "customer-centric" management focus.[100]

---

99 William Band, "The ROI of CRM Application Upgrades," *Forrester*, (March 2009).

100 Aite Group, LLC, "The Next Generation of CRM in Retail Banking: Sense-and-Response Marketing," June 2009.

Infrastructure improvements – Whether it is the latest Windows 7 upgrade, Oracle upgrade, cloud computing, virtualization, or another advancement, there are a number of technologies that can increase operating efficiency and provide increased processing power and flexibility. Institutions will increasingly look for infrastructure improvements that offer strong returns on investments.

Core processing technologies – As discussed earlier, perhaps the most important strategic enabler towards creating a more vibrant business model at a lower cost is a systematic review of the core processing area. Since many in the market today are still operating from some of the oldest technologies, it is no wonder that in the face of several new emerging areas, core processing replacement is high on their evaluation list.

"Through 2010, IDC Financial Insights expects bank IT executives to focus on fraud, risk management and core banking transformation as their key IT initiatives," states Karen Massey, senior analyst, consumer banking, IDC Financial Insights. "While 'rip and replace' strategies will not be prominent in the near future, investment will take the form of enhancements and upgrades to existing technologies, particularly while financial institutions continue to heal from the recession."

Planned IT Investments Through 2010

Q: Do you plan to invest in the following solutions in the next 12 months?

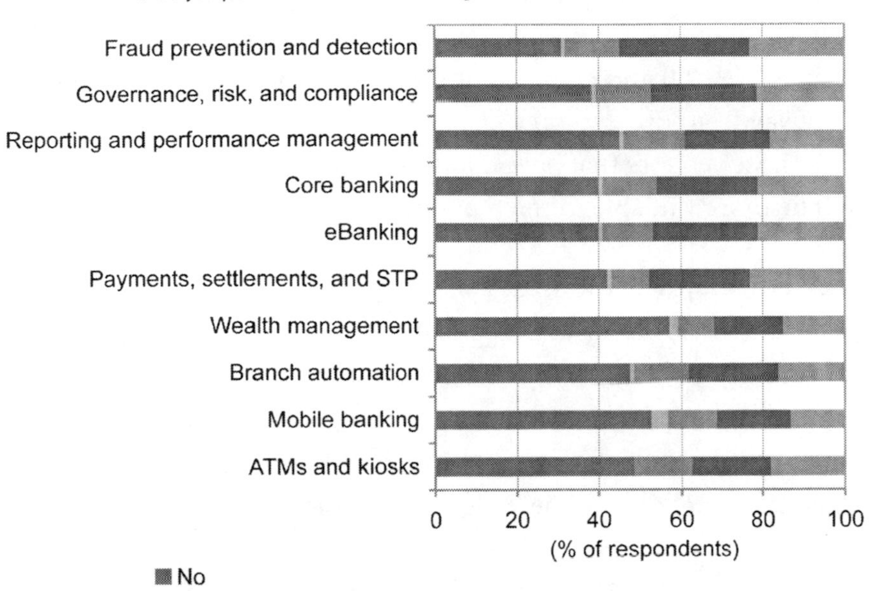

- No
- Yes, investing in a new solution for the first time
- Yes, investing in a new solution
- Yes, investing in enhancements or upgrades to the existing solution
- Yes, maintenance only of the existing solution

Source: IDC Financial Insights, 2009

Fraud prevention and detection as well as governance, risk and compliance will expect to see a spending increase through 2010. In fact, many respondents intend on investing in a new solution for these investments. Additionally, respondents also intend to spend on their core banking investments through 2010.[101]

Two significant trends may not result in technology spending but will have an impact on technology as an enabling strategy: vendor management and social media.

Vendor Management has taken on increased focus as financial institutions must ensure that their vendor contracts reflect their strategic plans. This area has also become critical to managing costs and meeting more stringent regulatory demands. Institutions must

101 IDC Financial Insights, "Business Strategy: IT Executives Survey Results – Can 2010 Get Here Soon Enough?", November 2009.

ensure that vendor strategies are aimed at benefitting not only the financial institution but the account holders as well. Vendors must keep the momentum on innovation and solution-specific advances, such as Web 2.0 and cloud computing, even as financial institutions shy away from unnecessary IT expenses. The economy will recover. And those vendors that have continued along the path of innovation will be more valuable in the marketplace.

Social media, also known as social networking, is a next generation (Web 2.0) set of tools and applications that allow people to create communities where they can share user-generated content including pictures, music, preferences, stories, and personal information with others easily and efficiently through a trusted network using a combination of old and new digital technologies.

Many boards and executives ask me what the social media phenomena means to them and how they should look at this growing area. In many respects, social medial simply reflects the changing demographics, globalization and interconnectedness that are part of our changing world. A new, low cost way to communicate, social media invites financial institutions to participate in a closer relationship with account holders who are willing to share an unusual amount of information about who they are, their preferences and their thinking. In return, they expect a more direct and open relationship with those who participate in the community.

A study by the American Marketing Association noted that 29.4 percent of respondents use social media to understand customer insights as a primary objective. In addition, 28 percent use social media to seek new growth areas.[102] Research clearly indicates that this isn't simply a fad, but a new era of how people interact with one another. The adoption of social media is widespread and continually growing.

---

102 Internet World Stats.

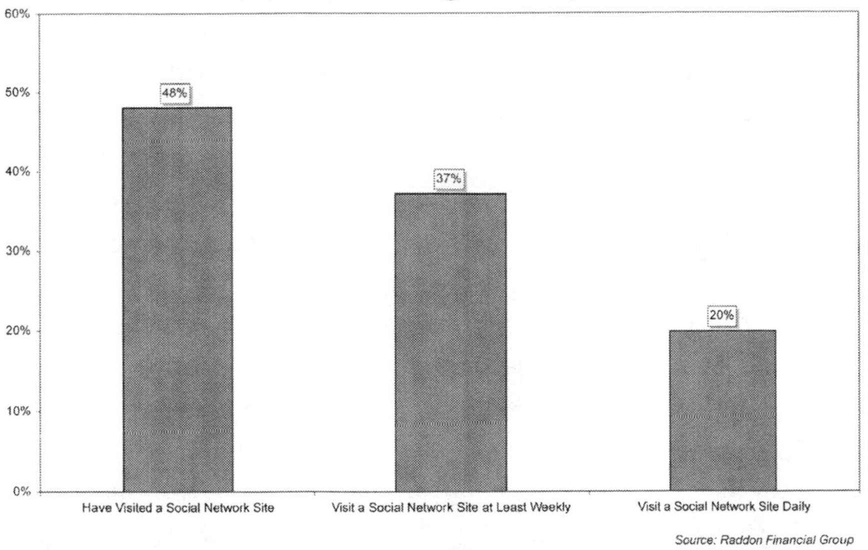

The growth of social networking can be seen in this chart. Almost half of all consumers indicate they have visited a social network site. Better than one in three visit a site at least weekly, and one in five visit at least one site daily.

Social media has continued to evolve rapidly. A decade ago, this technology, in its infancy, was only starting to evolve from Internet chat rooms and Web pages. Today, social media has become a force that businesses cannot afford to ignore. It provides consumers with an enormous amount of information and a platform for discussion to help them make better purchase decisions. These technological advances lead to free distribution, decentralized content creation, and increased non-branded information, which combine to enable the consumer to have much greater control. Sales and marketing teams are no longer the sole conduit for information. Companies find they must also rely on technology to determine customers' buying behaviors.[103] As banking moves to increased consumerization, this may also ultimately change the product development landscape.

103 Rycroft, "Technology-Based Globalization Indicators: The Centrality of Innovation Network Data," 2002.

Today, more than ever, consumers are able to speak their minds through various product review Web sites such as Epinions and cNet, social media including Facebook, MySpace, Twitter, LinkedIn, YouTube and other outlets like personal blogs. The digital convergence of all digital media – including mobile telephone, Internet, software and Internet platforms – is not only connecting people across the globe, but is also connecting businesses with their consumers.

Moreover, because technology now spreads these messages across the world in a matter of seconds, consumer opinion is receiving more attention from corporations than ever before. For example, CNNMoney reported on a Bank of America customer that had her interest rate "jacked up" to 30 percent. After her YouTube video rant received 350,000 views, the bank rolled back her rate. Unfortunately, user-generated content has no filtering process and there is no integrity check on what is said by individuals. At the same time, businesses are using this free and open dialog to push their own agenda.

For institutions focused on Generation Y, also referred to as Millennials, this can be an extremely powerful channel and tool. This demographic is becoming a powerful market segment. Over the next 10 years, the total income of GenY will reach about $3.48 trillion, according to Javelin Strategy & Research. GenY will begin to seek advisors they can trust, and will need to learn money management basics such as applying for loans or credit cards. Therefore, one of the challenges facing financial institutions is how to educate this next wave of account holders. Even more compelling is that this demographic is willing to share a tremendous amount of personal information, if you are a trusted member of their community.

Thus far, most financial institutions have yet to figure out their Web strategy when it comes to Generation Y. However, a few financial institutions are already onboard. For example, Stagecoach Island is an online virtual world created by Wells Fargo. Participants can explore the island and connect with friends. They can get virtual jobs, credit cards, and home loans. They can also learn about smart money management by visiting the Learning Lounge — a virtual Wells Fargo ATM.

But each geographic market segment may be different. GenY in Peoria could be quite different from GenY in San Francisco. That may mean creating a blog, Web chat or even a virtual world, like Wells Fargo. However, reaching GenY across the world will require a real cultural adjustment, not just lip service for the financial institution.

The payback from Generation Y is their allegiance. A survey of college students and graduates by Synergistics Research Corporation showed that 69 percent of graduates kept their checking account with the same institution they used in school.[104]

The new world "community" bank may not only refer to the financial institution in a neighborhood, but rather one where an affinitized group scattered nationwide, or even worldwide, connects regularly through social media. In fact, the "community" in banking has changed in many ways and many times over the years. Banks today offer a wider range of products and services than ever before; and they deliver them faster and more efficiently. Banking began as an endeavor around a geographic community or some affinitized group within a community. Today these social networking tools accelerate the ability to connect. Properly utilized, they offer the industry a chance to recapture their historic place as the trusted affinitized financial intermediary for communities. This is a significant opportunity if it fits in your long-term strategy.

---

104 Synergistics Research Corporation, "Stickiness Factor High With Student Financial Relationships," December 16, 2005.

4: Consumerization, Challenges and Brand

"It is time for a new generation of leadership to cope with new problems and new opportunities. For there is a new world to be won." - John Fitzgerald Kennedy

In 1997, a somewhat prescient New York Fed COO Ernest T. Patrikis warned of the commoditization of the industry in a speech before the Bank of Japan. "...this 'electronic comparison' shopping, that consumers will be undertaking in broader banking markets, will not necessarily always be for a 'complete bundle' of banking services from a single bank, but could be on a product-by-product basis, or, as financial products become increasingly standardized as they become automated, perhaps I should say on a 'commodity-by-commodity' basis," offered Patrikis. He went on to caution financial institutions of how the electronic delivery of banking services could erode the brand-name loyalty consumers now have to their banks. "Moreover, for products like credit cards, mortgages, auto loans, and some investment and savings products, there will also be nonbank competitors, perhaps "category killers" specializing in being a very large provider of just one product to gain the lowest unit costs through economies of scale in processing."

More than a decade later, the threat of commoditization within the financial services industry has led to greater consumerization of our industry. This is where the combination of the erosion of the mystique around the industry's products, better information about products and services, ease of comparison shopping, increased choices for consumers and efficient electronic access to electronic channels has allowed individuals to shop for financial products in the same way they do any consumer product.

The outcome is that pricing decisions by institutions and purchasing habits by consumers both begin to track traditional consumer buying behaviors. As a result, financial institutions will have to have more clarity than ever before about how they are different. They must increase their investment in their brand identity around this difference, and make the differentiation more obvious in every interaction with the individuals they are serving.

The Evolution of Consumerization

There was a time not too long ago when individuals conducted their financial business almost exclusively with one financial institution. It was during this period that financial institutions earned their status as trusted financial intermediaries by building their reputations based on trust, service and community affinity. While there are remnants of the original banking model that can still be found, it is becoming harder for consumers to distinguish meaningful differences between financial institutions. This jeopardizes the loyalty that once existed between consumers and their financial institutions throughout the industry. Institutions have been required to adjust to the changing landscape by trying to continue leveraging the investments in loyalty they have created in their own markets. They must balance this with the new realities of emerging market demands, changing product and service needs, a more transient market, demographic shifts, employee tenure changes, and other elements that have led to consumerization.

Consumerization has occurred as a result of six major trends:

- ***Industry has evolved*** – People previously saw the industry and its products as mystical entities that only the highly trained could understand. Today, more and more people feel comfortable with the basic product sets offered by the financial industry and do not see many meaningful differences.

- ***Product information is easier to obtain*** – Newer technology and social changes have allowed people to learn about and compare products, services and pricing quickly and easily. This makes it harder for institutions to differentiate and has driven down pricing.
- ***Access to product sources increased*** – It is much easier to get new products and services outside of an institution's immediate geographic area. Any loyalty tied simply to geography has eroded.
- ***Demographic shifts accelerated*** – Demographic changes have impacted the needs of the consumer and the way they view their relationship with their primary financial institution. Loyalty is being redefined and the formula for capturing a higher wallet share has changed.
- ***Retail differentiation attributes began to impact decisions*** – These trends led to consumerization whereby institutions have to work harder to show they are different and why consumers should do business with them. Their differentiation must become more obvious with every interaction.
- ***Some consumers see no difference between financial institutions*** – This is a factor wherein the threat of commoditization leads to consumerization. This translates into commoditized pricing for those who are unable to differentiate.

Other industries have already gone through this transition or have been consumer-oriented for some time. Even basic products that may seem highly commoditized, such as water, yogurt, coffee, soap, or even computer hardware, have attempted to differentiate themselves, and where they have not been able to, the pricing and margin results are dramatically lower. Consumers need help in distinguishing differences in all similar products in all market segments.

Exposure to Disintermediation

The trend towards consumerization is clearly evident in the evolution of services such as PayPal and Prosper.com, which blend social networking and banking. These services, in essence, directly link consumers in order to conduct traditional financial transactions such as payments and borrowing. These are also examples of how the industry, if it does not create an infrastructure that allows it to innovate at a lower cost, is in jeopardy of further disintermediation.

PayPal allows individuals and businesses to send electronic money online. It does not provide credit services, nor can users hold funds in their PayPal account. PayPal facilitates person-to-person payments (as well as person-to-business payments) but does not replace the financial intermediary, since an account with a traditional financial institution is required in order to either pay or receive funds from another individual. The payment process is simplified because payments are made to an email address. You can make a payment to any individual who has an email address through your laptop, home computer, or any mobile device with email capabilities – no check, debit card, or cash is required. PayPal's business model is to collect a small interchange charge paid by the funds receiver.

Because PayPal does not accept deposits, it is not considered to be a bank by the FDIC. However, a few states such as California and New York have raised questions about whether PayPal is operating an illegal banking service.

eBay recognized the compatibility of online payments with online auctions in 2002 when they acquired PayPal for $1.5 billion. The efficacy of this model is further evidenced by the fact that, only recently, a few banks have begun to private label PayPal capabilities. Also, as is the case with any successful business model, competitors are emerging. POPmoney by Cash Edge Inc. is private labeling a service similar to PayPal to financial institutions.

While services such as PayPal do not yet replace the role of a traditional financial institution, they are attempting to capture the higher value components of banking and capture the fees that have typically been garnered by traditional financial institutions. There is no reason why a financial institution should not have developed its own "PayPal like" system. But the creativity for such an action was stymied by inflexible systems and a lack of innovative cultures. While there are many innovative institutions with several unique products and services, this is an example of the value that is being transferred to new, more innovative companies as a result of inflexible, older internal systems and a lack of innovation. If others capture the higher value elements of the banking value chain, the industry will be left with the low value, truly commoditized areas of the business. This is an unattractive place to live and an impossible place to thrive.

Perhaps more directly, a potential concern of traditional financial institutions is the proliferation of Peer to Peer (P2P) lending sites doing business worldwide, including Prosper, Kiva, Zopa and LendingClub. P2P allows peer lenders and borrowers to find each other. Some of these sites include a traditional underwriting process, while others might instead verify the borrower's identity and allow lenders to review a portion of the borrower's credit information.

Prosper.com is an example of person-to-person lending without a traditional underwriting process. The dollar amounts range from $1,000 to $25,000, so the purposes of the loans will typically fall into what we would consider consumer and small business borrowing needs. All Prosper loans are three-year fixed rate loans. Investors can either invest into a portfolio of loans with a given risk rating (with higher returns for riskier loans) or can actually invest in individual loans. The matching of borrowers and investors (who are also individuals) is through a structured process, so the borrower's rate may be bid down.

Unlike PayPal or POPmoney, this process does indeed represent a replacement of traditional financial intermediaries. While participants are required to have bank accounts to facilitate the payment process, the bank or credit union role in the lending process itself is usurped. In some ways, this is how early banking began when individuals in communities would come together to help each other. This was actually the historic beginning model for credit unions as financial cooperatives. However, these newly developed programs are not in the not-for-profit model that credit unions were based upon. They are competitors for the profits of financial institutions.

How viable is this P2P lending business model? As of February 2010, the prosper.com Web site indicates that 29,000 loans worth $178 million have been generated since Prosper.com's inception in late 2005. The site also indicates that at this point 92 percent of loans are current. However, over 20 percent of the loans that the company originated since its inception have gone bad.[105] Clearly, the company has cleaned up its underwriting recently; however, prosper.com investors may still be carrying a very negative perception of this company.

Questions about the long-term viability of these models remain. If those with better lending profiles seek more traditional lending institutions and participate in a full underwriting process, losses may continue to rise for the higher risk profile individuals that many of these sites attract. Also, if regulatory concerns rise, the future may be limited to a niche role or become absorbed into a traditional regulated

105 Prosper, *Performance Data*, http://www.prosper.com/invest/performance.aspx..

financial institution. Few traditional financial institutions have a strong interest in a loan market where the average loan origination size is $6,000, like that of prosper.com. Most financial institution cost structures are too steep to allow this model to work.

Whether a non-branded or non-regulated lending entity will survive when losses begin to emerge, is yet to be seen. Individuals may begin to miss the value of the detailed underwriting process that has evolved in the banking industry over many years and borrowers may be reminded of the importance of knowing who their creditors are. Community based financial institutions demonstrated the benefits of the traditional model during this most recent crisis. While these new models might work well in good times when individuals can make loans with the security of those who could pick stocks during the Internet bubble or develop homes during the housing bubble, it is unclear of their survivability during more difficult times. It is easy to win when everyone is winning.

Both PayPal and P2P represent a signal to the market, if not an actual threat. The message is that if traditional institutions do not evolve to become more flexible and efficient, they will be increasingly disintermediated. They also demonstrate that as the market is consumerized, individuals and businesses will be willing to look for other niche players to meet their needs.

This is further evidence of why financial institutions need to take back their historic role as the primary, trusted financial intermediary in American communities – however large or small that community may be. To do this, they must embrace their regional competitive advantages while reflecting the demands of an increasingly connected and knowledgeable global marketplace.

Importance of Brand

This is a major reason why companies brand and position themselves to attract a specific audience or seek to establish an identity that distinguishes them from the competition. Nike Inc., for example, went from a $750 million business to a $4 billion business in just seven years in large part by capitalizing on its "Just Do It" branding campaign and the deep emotional link that people have with sports and fitness.

In the Interbrand Best Global Branding 2009 rankings,[106] the top five positions are almost predictable: Coca-Cola, IBM, Microsoft, GE and Nokia. Interbrand cited Coke for "showing the rest of the marketing community what it really means to manage a brand" with its campaign "Open Happiness," which targeted consumers longing for comfort and optimism in a tough time. Incidentally, the top ranked financial services company was American Express in the 22nd spot, and the top financial institution was HSBC (32nd), which is "performing fairly well, in part due to effective leveraging of online and self-service platforms and a growing exposure to emerging markets."

So how does the industry respond to its products becoming more and more of a commodity? The heart of financial institutions is fulfilling the role of a trusted financial intermediary. Traditionally a financial institution's role is defined as an organization that offers products and services that deal with, hold, invest and lend money.

Financial institutions should come to grips with the understanding that the profitable business model for buying and selling money is disappearing. The fundamentals of banking are changing. These changes have led to product diversification, a focus on non-interest

106 Interbrand, "Best Global Brands 2009 rankings," www.interbrand.com/best_global_brands.aspx.

fee income, a demand for greater flexibility and lower costs. All of this is occurring at the same time as the industry is commoditizing.

There are different ways to be successful, but any strategy should lead to redefining and then fulfilling the role as a trusted financial intermediary for your targeted market by delivering and demonstrating daily your unique value to the market. The financial institution should then ensure that it is culturally aligned with this differentiation strategy and invest in a brand that is associated with these strategic goals.

As an industry, financial institutions need to have more choice and more flexibility in preparing for what lies ahead. This fog of the future is very dense. Strategic plans that extend too far ahead are risky. We need to have a more responsive environment where we can react to market demands more quickly within the guideposts of our mission and values. A highly structured development process needs to be able to change quickly in the midst of this competitive environment.

Infrastructure to Win

In order to adjust to a consumerized market where you must be able to clearly demonstrate how you are different, having a flexible infrastructure in place that allows you to win is critical. This means focusing on all aspects of your business to create a more dynamic and flexible business environment to compete. This includes the people, processes and systems that you use to deliver value to your clientele.

As we enter this next phase of our more competitive industry, it will be critical to assemble a team that is not only experienced, but willing to acknowledge the fundamental pressures on the basic business model and that what might have worked 20 years ago may not be sustainable today. I often demonstrate to boards through a series of financial metrics that operating a financial institution today is more complicated and difficult financially than it was just 10 to 20 years ago. It's not because the products and services are so different, but that the operating model of flexibility was quite different a decade or two ago than what we are faced with today. Understanding how these products and services work and having a good knowledge of the overall industry is still very important. However, the team should recognize they have much less operating flexibility and that everything must become sharper, more directed and more innovative.

This should usher in a refreshed culture that embraces the strengths of the past and reinforces the importance of a performance-based culture that emphasizes an agile and creative environment that encourages innovation and results. Our industry has a reputation of being very measured and staid with good, stable and predictable careers. It was often viewed as less dynamic, but the trade-off for some was that it didn't require long hours (hence the old phrase "bankers' hours") and often led to a pleasant lifestyle. As a result of the changing dynamics, today's reality is that the level of effort, intensity of the competitive dynamic, operating challenges, and financial strains are as intense as any other industry. Any remaining remnants of a culture that doesn't recognize that things are quite different should be dispelled. We still must deliver on our role as trusted financial intermediaries in our communities, but our operating, efficiency and competitive standards have to match the new realities. These will have to be demonstrated not only through strategic positioning, but also through strategic delivery.

For example, assume that part of your renewed strategic focus is to launch products more quickly that meet the needs of your targeted demographic. Your team decides that by being able to tailor products, packaging and pricing more quickly, perhaps even at the point of service, you would reinforce your competitive difference to your target demographic. If you or your vendor are required to make the change in multiple layers every time you want to launch such a product package, even if it can be done, inefficiency will slow your time to market and create a form of tax on everything you do. The result is that you fall into the trap of lowered expectations and are back to where you started, always fighting an uphill battle. The risk today is not the perception of adopting more contemporary technology. The biggest risk would be staying with an operating infrastructure that will cause you to lose in the long run, even if your other competitive attributes have you in a good position today.

It is critical to do a strategic assessment of your infrastructure. I believe the incremental improvement approach has run its course. Instead, the industry requires a wholesale shift in how it uses and deploys technology in a way that allows innovation at a much lower cost and allows increased collaboration within the industry.

A management team can no longer be limited by their vendors or infrastructure, but instead must have the flexibility to meet changing needs quickly and at a low cost with the most contemporary tools available. This is about choice – choice on which hardware to use, choice on which operating system you can utilize, choice of what products and services you deploy (retail or commercial), choice on language or currency, choice on where in the world you want to operate, choice on pricing and bundling, choice to extend the applications on your own or through others, choice on which third parties you can work with – all from a single integrated application that is efficient. Ultimately, it should provide the management team the freedom to become the institution it would like to become.

Interestingly, technology has not taken away the desire for people to interact. Instead, technology has enhanced the human experience by automating lower level functions so that more time can be spent on the human interaction. These enterprise tools for the institution should allow human interaction to be more collaborative, personalized, and more powerful.

In order to find your difference and deliver it efficiently, it will be important to have an infrastructure that allows you to react to the changing market; limited only by your own imagination and the capital and talent of your staff.

In this way the paradigm between the vendor and financial institution must also change. The historical approach of a vendor building products for the marketplace based upon a handful of discussions with the hope they get it right will have to give way to a more collaborative "real time" approach. This means all products will be created in a way that seeks solutions that work for all and, in doing so, grays the line between vendor and partner. A model, where real time interaction between all parties in a technology decision will result in more powerful partnerships for the long-term, has a greater strategic impact on the business.

The processes deployed with the right people and systems are also very important. Once a financial institution has clarified how it will compete in the long run, identified its competitive differences, assembled the right team, and refreshed its operating infrastructure, it must then review all of its key processes. A leaner, more directed cost structure around your key competitive differences is part of the cultural shaping that should take place. Besides eliminating products and services that don't fall into your core competency, all processes that make you less efficient and take away from your strategic delivery to the account holder should be modified and improved.

Just like the technology infrastructure, the old model of small incremental improvements will no longer be sustainable because there is too much pressure on the operating model and not enough time given to the competitive dynamic. Change for the sake of change is not what we are after. Instead, it is a necessity to have an approach that allows more creativity for delivering value to account holders by spending in areas that have the greatest impact to the consumer while streamlining all lower-level or lower-value functions.

This is another reason why the technology environment is so important. I sometimes hear that the functionality of an older legacy system is why an institution maintains a series of older platforms. What they don't factor is that with a more contemporary architecture, it is much easier to add functionality than with older systems. These older systems have often been around for decades and naturally have hard-coded functionality built in over that extended period. But they make it more difficult to deliver on all the most important strategic challenges today, which is why they must be addressed. Strategic decisions around your technology are likely the most important strategic delivery enablers as they provide the tools your team will need to improve delivery and drive costs lower.

With Change Comes Opportunity

Adapting to new environments is something banking has been required to do from the beginning. In early civilizations – such as Egypt and Mesopotamia – temples stored the gold for safekeeping. They were the trusted intermediaries of their day. However, the gold remained there unused, while others in the trading community or in government had desperate need of it. In Babylon, at the time of Hammurabi in the 18th century B.C., there are records of loans made by the temple priests. The process had evolved from the storage of gold to the lending of it for a return plus some additional gold for the privilege. The concept of banking had arrived. [107]

---

107 Bamber Gascoigne, "The empire of Hammurabi: 18th century BC," *HistoryWorld*, http://www.historyworld.net/wrldhis/PlainTextHistories.asp?historyid=aa10#86.

Banking activities in Greece became even more diverse and sophisticated with private entrepreneurs, as well as temples and public bodies, now carrying out financial transactions. They took deposits, made loans, arranged for credit, exchanged money, and tested coins for weight and purity.

Rome, with its genius for administration, adopted and formulized the banking practices of Greece. By the 2nd century A.D., paying an appropriate sum into a bank officially discharged a debt. Appointed public notaries registered such transactions.

The collapse of trade after the fall of the Roman Empire made bankers less necessary than before, and hostility of the Christian church toward the charging of interest hastened their demise. With that, usury came to seem morally offensive so banks went from being trusted intermediaries to persona non grata.

Banking has gone through a number of makeovers since then, but today it is under fire again with the trust and loyalty eroded greatly, particularly among large financial institutions, and significant questions about the future model for basic financial institution product offerings. The credit crisis and subsequent aftermath has stirred an active debate about the appropriateness and risks of the integration of different types of products under the "one stop shopping" concept, the role and failures of regulators and government oversight, public policy and its impact on the financial services market, and how best to create an environment to protect consumers and businesses while at the same time stimulating economic growth and vitality.

In the credit union space, it has also spurred an intense debate about the corporate credit union network and its viability and future role in the industry, especially after having caused significant pain through its investment losses in an industry segment that was otherwise well-capitalized and well-positioned from a risk profile perspective. The corporate credit union failures and subsequent financial impact has altered the industry dramatically.

In terms of reputational risk, many of the larger institutions were more actively involved in the expansion of the basic banking business model and have therefore suffered a disproportionate amount of reputational "fall-out" from the economic downturn. This is, no doubt, influenced by the dramatic headlines, government bailouts, populist rhetoric and real animosity over items such as the timing of management bonus payments.

While community-based institutions largely did not participate in some of these riskier expansion areas, they were still impacted by the overall economic slowdown, particularly in housing and auto loan defaults, and increasingly in small business or commercial real estate lending.

The recent reputational damage to many of the country's largest banks has created an opportunity for all community-based financial institutions, banks and credit unions alike, to demonstrate their value and capture share.

However, as evidenced by recent research conducted by Raddon Financial Group, it seems that this reputational damage might be an opportunity for institutions to recapture market share if they can convince the public of their ongoing strong commitment to the historical tenants of the industry while driving better service and lower costs. For now, community-based institutions, in particular, may have a unique opportunity to leverage the market perceptions. However, they too will have to address longer-term business model issues even if they continue to see gains in their share of deposits and lending.

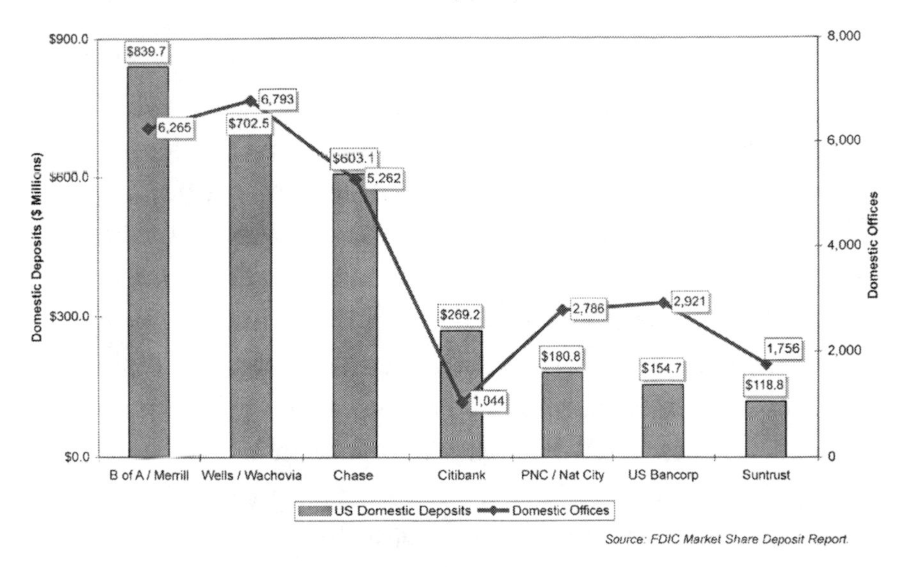

The U.S. banking market is increasingly dominated by three large players: Bank of America, Wells, and Chase. Paradoxically, this represents an opportunity for the community-based organization that is able to demonstrate its knowledge of and commitment to its local communities.

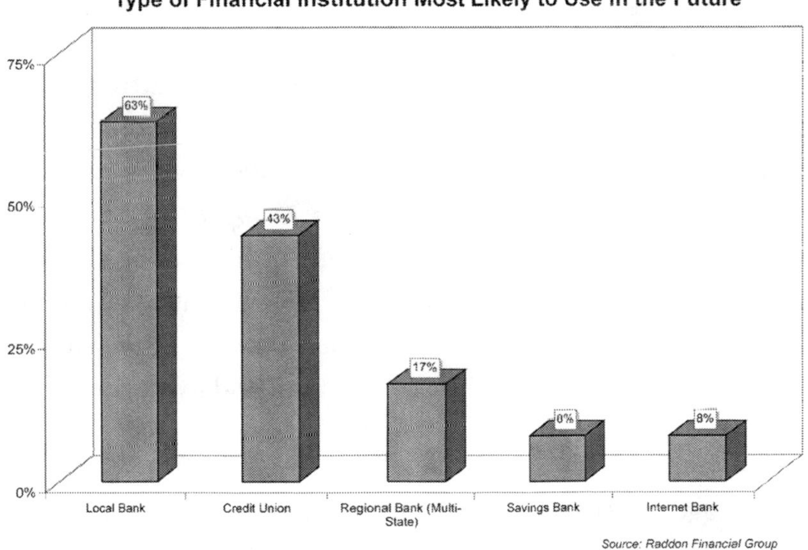

Evidence of U.S. consumers "thinking local" is seen in this research conducted in 2009. Based on what had happened to the banking industry in 2008 and 2009, consumers expressed a strong desire to establish relationships with local banks and credit unions.

Find Your Difference

I have the opportunity to participate in many strategic planning sessions for financial institutions. From that experience, I find remarkable similarities in the basic strategies and challenges. Most financial institutions have a set of strategies aimed around service, trust, pricing and community affinity. The difference I find between the discussions at the planning sessions and the resulting performance is the cultural commitment to make their differentiation point more obvious and reflected in everything they do. Naturally, capital and scale can limit how much can be pursued in a prescribed period of time; however, today's global environment offers boards and management teams a unique opportunity for greater clarity about how they compete. Rather than chasing basis points on products they often fail to understand or lack sufficient expertise to make the cornerstone of their margins, they should instead, embrace more completely those particular products and services where they can really demonstrate a difference - *finding and creating market niches where they can win.*

It is crucially important for each institution to be extremely clear about what it has defined as its primary competitive advantage. Michael E. Porter, a recognized leader in competitive leadership and Harvard Professor, in his 1985 book *From Competitive Advantage to Corporate Strategy*, introduced the concept of the value chain. The book describes the two prime elements of competitive advantage – differentiation and cost leader – and the various mechanism of the value chain that permit for the development and execution of these strategies.[108]

Porter suggested that activities within the organization add worth to the service and products that the organization produces, and all

108 Michael E. Porter, "From Competitive Advantage to Corporate Strategy," *Harvard Business Review* (1985).

these activities should operate at optimal level if the organization is to achieve any real competitive advantage. If operated efficiently, the value obtained should exceed the costs of running them. For financial institutions, this translates to account holders retaining, or returning to, the organization and transacting regularly. Porter suggested that 'primary activities' and 'support activities' are at the center of this strategy. Primary activities consist of inbound logistics, operations, outbound logistics, marketing, and sales and services. The support activities help the primary activities in helping the organization attain its competitive advantage. They include procurement, technology development, human resource management, and firm infrastructure.

Other industries have addressed an otherwise commoditized market by consumerizing and differentiating. Some more familiar examples of brand affinity include Volvo (safety), Nike (courage), GE (imagination), Walmart (value), 3M (innovation), Virgin (anti-establishment), and Honda (dependability).

Making a small change to a commodity that fits a need in consumers' lives can be very profitable. Take, for example, bagged lettuce – a bag of cut and washed lettuce costs a multiple of what a similar-sized head of lettuce costs at the supermarket. Fresh Express created the first ready-to-eat pre-packaged salad in 1989. It was an opportune time to launch such a product, given the emphasis being placed on convenience in two-income households and the growing nationwide focus on eating healthier meals. Who would have guessed that repackaging an otherwise commoditized set of products (lettuce, carrots, etc.) in a convenient ready to make bag would have such an appeal? Fresh Express did so by addressing a specific niche, not to those who wanted to prepare healthy alternatives at home, but to those who wanted to serve a healthy alternative easily and at a lower price than the same item would cost at a restaurant. Fresh Express increased revenues almost six fold between 1991 and 1994, with the retail business leading the way. Chiquita purchased Fresh Express for $850 million in 2005. Today, the market for this product category is now over $3 billion in the U.S. alone.

Another example is the iPhone. In June 2007, Apple entered the crowded smart phone market with a price premium product and limited distribution (only one wireless carrier); but they offered differentiation through innovative design, superior user interface and an improved interactive display. Almost none of the components or software was unique within itself, but the combination of these elements was extremely appealing in the marketplace. In less than a year, the iPhone captured more than 20 percent of smart phone market share, selling 4 million wireless devices per quarter. The market has rewarded Apple for its innovation. Between January 29, 2007 and January 29, 2010, Apple stock rose 123 percent, while Motorola's stock price declined by 66 percent in that same period. Motorola, which continued to add features to its more outdated base technology for smart phones, has announced plans to spin off its mobile phone division.

In the case of iPhone owners, they now have a choice of almost 100,000 applications designed specifically for the iPhone, from games (entertainment makes up one-third of the choices) to more than 1,000 personal finance applications that range from remote banking applications to simple bill reminders and checkbook registers. Unlike other cell phone users, 72 percent of iPhone owners download five or more applications. Consequently, it is not surprising that some inventive financial institutions are creating iPhone applications for mobile banking.

Firms that were swift to respond with iPhone applications— like Bank of America, Grupo Banco Popular, and USAA — have announced that iPhone users already account for approximately 40 percent of their mobile banking users, according to Forrester.[109]

Niche Players in the Financial Services Industries

While many national, regional, and community financial institutions try to be all things to all people, some institutions have carved out a unique niche in the marketplace. For instance, Northern Trust Corporation, which has been in operation for 120 years, has grown to $82 billion in banking assets by delivering investment management, asset and fund administration, fiduciary and banking solutions specifically tailored to corporations, institutions, and affluent individuals. For its non-traditional banking efforts, as of December 31, 2009, the company had $3.7 trillion in assets under custody and $627.2 billion in assets under management.

More recently, ING DIRECT has had a meteoric rise in the financial services industry by offering a commodity savings product that is high-volume and low-margin. Eschewing traditional branches for only several retail stores (a.k.a., ING Cafés), the firm's business model requires that consumers open and maintain their accounts electronically. As a result, their product delivery system appeals to those consumers who are more technology oriented and do not feel they need a lot of hand-holding or empathetic contact. Since 2000, the company has grown to $74.3 billion in deposits.

Evangelical Christian Credit Union (ECCU) has grown to $1.2 billion in assets and $821.7 million in deposits since its inception in 1964. Uniquely, this California-based cooperative has become the leading banking resource for growing churches, Christian schools, and other evangelical ministries throughout the United States. In addition, missionaries in more than 100 countries also avail themselves of

---

109 Emmett Higdon and Alexander Hesse, "iPhone Apps Fill Mobile Banking Gaps," *Forrester*, (September 2009).

the services provided by the credit union. To meet the needs of its constituents, ECCU has a wide range of financial services specially designed to make evangelical ministries more effective. For example, the cooperative offers financing options that minimize distractions from ministry and cash management resources that are tailored to the unique needs of churches, schools, and mission-sending agencies.

Following a theological theme of its own, University Bank in Ann Arbor, Michigan, is catering to the Muslim community by offering Shari'ah compliant financial products (deposits, loans, insurance, and mutual funds). By doing so, the firm hopes to become the single point of contract for followers of the Koran.

A number of other select American financial institutions have also made and are continuing to make their mark by serving various minority and ethnic communities. For instance, Carver Federal Savings Bank in New York City, founded in 1948, has become one of the largest African-American operated banks in the United States with $791.4 million in assets and $603.4 million in deposits. The savings institution specializes in serving African American communities whose residents, businesses and institutions have had limited access to mainstream financial services. In another instance, Cathay Bank began operation in 1962 for the sole purpose of providing financial services to the growing but underserved Chinese American community in the greater Los Angeles area, thus becoming the first Chinese-American bank in Southern California. Having grown to $11.5 billion in assets and $7.5 billion in deposits, the firm's service network now extends from the West Coast to New York and Massachusetts on the Eastern Seaboard, north to the state of Washington, and south all the way to Texas. The company contends that its understanding of the culture and business practices of Asia, its vast experience in trade finance, and its extensive correspondent bank relationships, will make it the clear bank of choice for companies doing business between America and Asia.

In more recent times, there has also been the emergence of a number of institutions that focus on serving the Hispanic community. For instance, Solera National Bank, established in 2007, set the standard for community banking and serves the Hispanic community by providing financial education and leadership. Accordingly, the bank is not only offering traditional bank services, but also remittance and check cashing services. Although striving to close the majority achievement gap, the Colorado-based firm is committed to serving the diverse consumer, professional and small-business markets in the Mile-High state. In another instance, Libertad Bank was formed to cater to the Hispanic market in Texas. Uniquely, the firm is a hybrid community bank and check casher (i.e., check cashing, Western Union money transfers, and prepaid phone cards), which are services that most Hispanics want and need, but which most banks do not provide.

Here are examples of how other financial institutions have also attempted to establish differentiation niches for themselves:

- **Keypoint Credit Union** (Santa Clara, Calif.) was one of the first financial institutions to use Facebook. Keypoint leverages social networking by offering access to account information and credit union news; prospective clients see offers and instructions on how to join; attracting technologically-advanced mobile consumers; and running sweepstakes promotions to attract account holders' use of the new channel.
- **Susquehanna Bank** (Lancaster, Penn.) is trying to develop a slot for nonprofits. Nonprofits have become a central part of its growth strategy. They have assembled a bundle of banking products for nonprofit organizations, including discounted fees on services such as checking accounts and remote deposits. In 2008, the financial institution offered $100,000 lines of credit for volunteer fire companies. It also

offered free educational programs for nonprofit executive directors on topics ranging from grant writing to changes in the 990-tax form. Susquehanna Bank generated more than $10 million in deposits from nonprofits as a result.

- **Ixe Banco, part of Ixe Grupo Financiero SA De CV of Mexico**, is a group of financial service businesses created in 1994. Ixe Banco, which caters to large tier, high-end customers, decided that an open plan environment and automated dispensing of cash would allow tellers to offer a more proficient and interpersonal customer service, as well as boost cross-selling opportunities and improve branch sales performance. The bank complemented the cash dispensers with a coffee bar, Internet-enabled PCs and a children's play area, all for use by customers while branch staff completed their banking transactions. Customer satisfaction is reflected in the bank's rapid growth. Ixe Banco's initial Pacifico branch grew by $80 million and won almost 400 new customers in six months following the launch of the new concept in 2005.

Considerations in finding your forte include: discovering how you are different, understanding the demographics you are servicing, creating a focused strategy encompassing those elements, making sure you have the personnel to execute the strategies, and maintaining a motivated and qualified leadership team, thus ensuring that everything you do emphasizes how you are different.

Account Holders Looking for a Reason to Smile

With the declining customer satisfaction rates in the industry, it is a very good time to find your niche, make it more obvious and win market share.

J.D. Power reports in its 2009 Retail Bank Satisfaction Study that, "driven by declines in both satisfaction and brand image, customer commitment to retail financial institutions has decreased in 2009." This marks a steady decline in customer commitment since 2007.

The study reports that only 35 percent of customers said they were highly committed to their retail bank in 2009, compared with 37 percent in 2008 and 41 percent in 2007. On average, highly committed customers use more products, give more referrals and are much less likely to switch to another bank, compared with customers who have lower commitment levels.[110]

Rather than finding ways to deter clients from leaving, industry leadership has allowed a slow steady decline in loyalty. Instead of meeting their clients' needs, leaders have permitted old nagging issues, such as technology, to linger in a deficient capacity. Legacy technology provides a severe competitive disadvantage that the financial industry must address quickly because it is at risk for the medium term.

"Customers reporting the lowest levels of commitment in 2009 happen to be those with deposit balances that are 15 percent higher than average," said Michael Beird, director of the banking practice at J.D. Power and Associates. "With this in mind, it is crucial that banks take steps to address this steady decline in customer commitment, as moving just 5 percent of customers from low and moderate levels of commitment to high commitment can mean additional deposit growth of more than 2 percentage points higher than average. This is critical in an environment where 4 to 5 percent is the norm."

---

110 J.D. Power and Associates, "Customer Commitment to Retail Banks Declines for a Second Consecutive Year," May 19, 2009, http://businesscenter.jdpower.com/news/pressrelease.aspx?ID=2009087.

Even credit unions that have enjoyed an uptick in new members and deposits face the challenge of loyalty drain and reducing market share as the primary financial institution for individuals.

The macro global trends punctuated by the recent credit crisis have energized financial institution leaders by creating a willingness to take a renewed look at their business. What they'll find is a business model that has been under pressure for more than a decade and a realization that the threat of commoditization has resulted in the consumerization of our industry. This has manifested itself in a variety of ways, including eroding loyalty and less operating flexibility. This puts more pressure on financial institutions to reassess how they are different; whether their people, processes and systems will enable them to deliver this difference in a more compelling way; to eliminate all areas of the business that don't support their competitive positioning; and invest in a brand that reinforces their strategy.

The financial crisis has resulted in a concentration of the financial industry around a few very large banks, which, for a variety of reasons, have suffered reputational damage. Community-based institutions willing to address these underlying business issues and be bolder about their future have a chance to not only survive, but thrive by providing the roadmap to a profitable, service-oriented model that can recapture the imagination of the industry and provide the blueprint to the future.

5: How to Succeed: A Call to Action

"It is time for us all to stand and cheer for the doer, the achiever – the one who recognizes the challenges and does something about it." – Vince Lombardi

---

The financial services industry has experienced a steady drumbeat of change over the past two decades culminating in the more dramatic headline-grabbing events of the recent credit crisis and its fall-out. As signs of stability emerge, the industry is left to reflect on the lessons learned and face the reality of a very different operating environment. It has also allowed many leaders to finally acknowledge that there are more serious underlying issues that must be addressed in the basic business model for banking. Many of these changes have set the table for a new model in banking that will have to reflect the economic, geopolitical, technological, and demographic trends, as well as industry specific issues of margin compression, regulatory pressures, cost concerns and consumerization. Financial institutions should now incorporate the lessons of the past and the recent trends, and begin to craft a differentiation strategy for moving forward.

In the process, we'll have to set a course to recapture the role of the primary financial intermediary for the communities we serve in a way that allows us to serve our constituents better and more efficiently. The good news is that people and industries have had to adjust to dramatic changes before.

Take for example, Caterpillar, which came close to bankruptcy in the early 1980s as they became stagnant, bloated and unable to respond to the intense competitive pressure applied by its Japanese rival Komatsu (who at the time used the internal slogan "encircle Caterpillar"). Much like our industry, Caterpillar was at a crossroads and represented a proud history in the annals of American business. But, it was on the brink of failure. Caterpillar responded by reengineering the entire organization from process improvements to more flexible production facilities and a more flexible workforce. They became a more agile manufacturer with a refreshed product suite and much leaner production processes. It was a difficult time, but as a result of its willingness to make wholesale changes to its operations and culture, Caterpillar reasserted itself as the dominant competitor in heavy machine equipment around the world. This openness to adopt change in the face of serious market challenges has resulted in the company being able to capitalize on the globalization of its industry. In 2008, Caterpillar Inc. achieved record sales as booming demand for construction and mining equipment in China and other emerging economies offset weakness in the U.S. market.

Our industry is in a similar situation. Our basic business model has been steadily eroding for decades, the fee income used to prop up our model is under attack, and the complexity and costs have increased in a variety areas. We must also deal with a backdrop of fierce competition, a changing human landscape that shapes needs and buying behaviors, channel proliferation where users want to access their balances more frequently through more conduits, an intensifying regulatory environment, heightened fraud and security risks, and now the industry reputational damage from the global financial crisis. This has forced a number of boards and executives to be willing to think differently about how they compete with renewed openness, just as Caterpillar had.

The industry also began to see increased consolidation as well as the threat of commoditization. This led to the consumerization of our market. The result was that some in the industry began chasing basis points on products they did not understand and increasingly relied on the resultant fees they generated.

At the same time, we have some of the oldest and most inflexible technologies of any industry. Making it worse is that layers of newer technology were designed to extend the life and disguise the weaknesses of these older platforms. The result is an expensive and inflexible environment that the industry began to accept as normal. Unfortunately, this known, long-standing issue has become intolerable in today's marketplace.

The result is that most institutions are looking to diversify revenue streams by adding products and services that meet their niche and consumer demographic more quickly. They are looking to increase non-interest, fee-income sources to blunt the impact of margin compression, to know more about the consumers they service while getting to know them better and thereby deepening their relationships and increasing their ability to cross-sell other products and services tied to their brand.

Financial institutions would also like to increase their ability to meet regulatory changes more quickly and efficiently. In addition, institutions need to find a way to differentiate themselves more clearly. While they are doing this, they need to create a more flexible infrastructure that allows them to meet these demands quickly while providing this flexibility to adjust to market demands. All this while they are being forced to create a lower cost infrastructure and increase innovation. This is causing many to abandon the old formula of small incremental changes that is simply no longer good enough. Institutions have to take a more serious top-to-bottom review of all aspects of what they do, their processes, products and people to set a new, bolder and clearer path for the future.

Back To Basics

Following the recent whirlwind of change, it is time to turn back to the basic tenants that made this industry so successful. We must change almost every facet of how we pursue our basic tenants of existence – *trust, service and community affinity.*

For financial institutions to emerge as leaders again, we need to focus on what we do best, leverage our strengths, apply traditional business metrics to decisions, and create an infrastructure to compete and win. All decisions should leverage a financial institution's key differentiators and make these differences more obvious. At the same time, they must emphasize a rededication to a higher level of service and convenience, building efficient, trusted relationships, accentuating their position in the community, and fulfilling their historic role as trusted financial intermediaries.

In the process, we must work with our political leaders and global counterparts to create a regulatory and legislative framework that is more effective, eliminates waste and competing agencies, while providing a safe and sound financial system that the industry can be proud of and the account holders can trust.

Community-based financial institutions have an opportunity to become the inspirational leaders in a reenergized global economy. As financial institutions, they can lead the worldwide financial recovery by recapturing their status as trusted intermediaries starting with reestablishing their reputations one community at a time. In the process, they'll need to reshape the business model that allows them to serve their community better and much more efficiently. These efficiency and flexibility gains can be used to fund service levels, product targeting and pricing and packaging, among other strategic areas of focus.

This network of community-based financial institutions focused first on servicing local needs in the face of a changing global backdrop may be the most important tool in the economic recovery. These small to mid-sized institutions may be viewed as too small to matter in a world of "Too Big to Fail," but these same institutions may be too important to ignore or take for granted.

Community-based institutions have the tools and are in the perfect position because they have the potential to be the key economic engines in their communities. For each financial institution, getting there takes some calculated planning and answering some key strategic questions:

- What are our sustainable differences compared to our competition?
- How do we make our distinction(s) more discernible with every human interaction?
- Is our organizational culture unified around these qualities?
- Is our cost structure targeted around marketing, delivering and supporting these differences?
- Do we have the right people, processes and systems to execute our differentiation strategy?
- How do we use our differences to fulfill our role as a trusted financial intermediary?
- Are we willing to commit to achieving a leadership role?

A summary checklist of some key steps to help navigate this process might include:

- ❑ Establish what your sustainable competitive advantage will be based on a select set of competencies that will help you meet your targeted market demands.
- ❑ Assess the macro environment and its impact on your particular targeted market and demographic.
- ❑ Inventory all your products and services. Be brutally honest about what your core competencies are.

- ❑ Create a technology infrastructure that is flexible, proven, open and cost effective.
- ❑ Ensure the proper infrastructure to limit exposure to regulatory, security and fraud issues efficiently.
- ❑ Develop fluid strategies that adapt to change quickly, get closer to the person(s) you are servicing, and include them in decisions about what to offer.
- ❑ Create a mechanism to learn everything you can about the people you serve in a way that allows you to use the information to serve them better and more efficiently.
- ❑ Ensure that you are consistently treating a person the same way through all channels in terms of pricing and promotion based upon their profile.
- ❑ Collaborate with others in the industry on non-strategic core processes to drive down costs and increase innovation.

"By focusing on aspects most critical to the banking experience, financial institutions can win the favor of their customers, which can lead to considerable financial rewards," said Michael Beird, director of the banking practice at J.D. Power and Associates.[111]

The culture of your technology partners is critical as well. Certainly, no partner is perfect. However, finding partners that do not limit you and also provide you with the tools and cultural commitment that is crucial to your success, is imperative. It could mean the difference between success and failure. If your technology vendor is not investing, they are milking recurring revenue and not investing in their solutions. This not only hurts you as a partner, but also makes their business non-viable in the long run. Even though they might not always agree on everything, as no relationship is perfect, business partners need to share a common view of how to succeed

111 J.D. Power and Associates, "Customer Commitment to Retail Banks Declines for a Second Consecutive Year," 2009.

and how a relationship should work. Their vision for the future and investments in areas that will benefit you and the industry for the long term are critically important. These factors could improve the financial performance of your financial institution and assure that you have an adequate return on investment.

In addition, you can address the heart of your technological infrastructure by examining your core enterprise software, which is often the weakest technological and performance link. The right core system can eliminate layers of applications and redundant disjointed information silos. It can also re-engineer processes across the organization. The right choice in your core enterprise software will likely afford you the freedom to respond more quickly and invest in the areas that are strategically more important to your institution's success and vitality.

In the final analysis, financial services executives will have to establish where they want to direct their institution in light of the gathering pressure of changing market forces.

Innovation and Leadership

The time is now for a few good leaders to show the way. It seems like there is a never-ending list of challenges that we face. Life is becoming more complex, the world increasingly global, and the rate of change is accelerating. From economic woes, educational system shortfalls, environmental concerns and healthcare costs, to family issues and gas prices we have much to occupy our minds. In the business world, one thing that has cut through the daily wave of concerns is the steady drumbeat that represents the drive towards innovation and leadership.

I have found that one of the great attributes of capitalism, in all its forms, is that it was born out of a sense of rugged individualism where everyone is capable of standing on their own two feet, relying on their wits and willing to innovate and make a go of whatever was important to them. Periodic setbacks are part of the path to grander and more impactful success. The kind that stirs the imagination and excites the soul. It is socially acceptable to drive to succeed through measured risk and a desire to improve and show a new and better path to all aspects of how we live. This kind of culture invites innovation and leadership.

In the face of an increasingly fast-paced world, when oil prices, Internet bubbles and housing prices seem to change even more rapidly, what role does each of us have? More specifically, in the face of a rapidly changing landscape for financial services, how will we find the path to success? Who will be willing to show the way to a better tomorrow?

History has shown that the centerpiece of true value creation begins with a willingness to innovate and to lead. This fearless quest comes with plenty of risk, but none greater than stagnation and ultimately a mediocre existence. An existence that so many will find themselves in as they rationalize away why all the paths to breakthrough performance can't be done, shouldn't be done and won't be done.

It is precisely at times like these that we need to lean more heavily on innovation and leadership as guide posts in an uncertain world. They can lead us to that magical place where personal satisfaction, economic achievement, and positive social impact come together like a fine symphony.

Now more than ever, we need individuals and groups who are willing to not only help reassert the role of financial institutions as a key ingredient to economic vitality, but also inspire our communities and in the process drive the global economic recovery, one community at a time.

Plan of Action

I have outlined below some of the attributes that I think will be necessary for this to happen. This is by no means a complete or perfect checklist. It is a set of guidelines that may compel you to action. It is a proposal for a renewed commitment to embrace your opportunity, as a member of this community of financial institutions, to make a difference. It is an industry you know well. Recommitting to a purpose of inspiring those in your affinity group to work collaboratively, helping your community one relationship at time, and motivating people in your institution toward a renewed vision will provide the impetus for economic vitality in your region.

Here is the plan:

✓ ***Set your standard*** – A key step for this next phase is setting a personal standard of commitment and performance. A personal decision made by each of us demonstrates that we are willing to commit to what it will take to influence our institutions, our industry and the communities we serve. This begins with a clear view of what our goals are, what the desired future state looks like, and by making a commitment to ourselves as well as our team, family and community on the standard that we hold ourselves to and why it's so important to support this commitment. Only by clearly and overtly making this personal commitment to a standard of performance, can we create the energy and perform at a level necessary to change our communities and the industry.

✓ ***Find your passion*** – So many of us weigh the risks and rewards of any endeavor, sometimes to the exclusion of the best opportunities. Pursuing what you are passionate about and enjoying the journey carries almost no risk because the likelihood of success is higher and the self-satisfaction becomes self-fulfilling. People often ask me if they should do what they are good at or what they make money doing. I say move to where your energy is, because ultimately your hierarchy of needs is met through passionate performance of what you love to do and the satisfaction of knowing you had an impact. Make sure you are clear on what you love about the industry and how you want to influence it. While we will always measure the risk, there is no risk in pursuing what we love to do. Great people always find a way to succeed when they are passionate about making a difference. Passion is usually governed by values, such as wanting to make a difference, and drives productive behaviors. Recapture what you love about this industry and set out to make a difference.

✓ ***Find the best people*** – This is about the oldest cliché we hear– the importance of people. And it still holds true. It is always about the people. It is important to find people who share a common view of how you are going to be successful. I find that most successful people naturally self-select others who share the same standard for performance, passion for a goal, and desire to succeed. One of the hardest things to do in life is to convince qualified, high performance people to do what you want them to do because they want to do it. That means you have to start with the best people, understand their needs, and find a path for them to follow whereby their success means success for everyone involved. Entrepreneurs who learn to do this find multiple successes because the best people want to follow them. It's also why some investors have successful executives who want to make money for them and others who don't. We must have the best teams leading the next phase of this industry's transformation. Get yours together and get to work.

✓ ***Trust your team*** – Once you find the team that will help navigate the future, you have to trust their commitment to the team's goals, their abilities to make good decisions, willingness to seek council when needed, be passionate about their work and always have the team's goals as their first priority. The task ahead cannot be done by individuals, but by groups and teams who share a common vision for the future, are committed to each other, and have a passion for succeeding in every endeavor.

✓ ***Become the ideal student*** – If you know this is important to you, you must invest the time to know everything you can know about the industry, the competition, how to succeed, your clientele, and your stakeholders. Knowing what you want and who you are is essential as you create your self-image and pursue it. It is very important to be clear about the person you want to become, and make every effort, every decision and every interaction toward pursuing that image. Be prepared for the battle ahead by knowing your industry in all facets.

✓ ***No short cuts*** – I wish there really were short cuts to success – like the 10-minute workout, the easy money, and the performance without the work. Unfortunately, any great success requires sacrifice, prioritization, and dedication to work hard. This does not mean that you should not work smart and be as efficient and innovative in your work as possible. Nevertheless, we must make a commitment to put in the necessary work. We have to be as clear as we possibly can on the standards we are setting for ourselves and devote the passion and energy required for success. Create good habits. The mind is always looking for habits as a way of avoiding new decisions. For example, putting your keys in the same place every day or getting into routines allows us to free up time and energy on the more difficult and energy sapping areas of new decisions. Therefore, it's important for these habits to be positive, such as good eating habits, getting a good night's sleep, making your work day a priority, and creating a habit of thanking those who are committed to common goals. There are no short cuts and you must create good productive habits for success.

✓ ***Be realistic*** – We must be realistic about what is possible for people. When you are knowledgeable about an industry, are passionate about changing things for the better and have the commitment to make a difference, you have many of the attributes for success. We must also be realistic about the cost of making success happen, its risks and the amount of effort required to pursue the vision. I define an optimist as a person who sees the brutal facts and ignores them in an effort to achieve their goal. A pessimist is the opposite. They see the brutal facts and just quit because they can't imagine a way to overcome the challenges. So you have to be a realist, a person who sees the brutal facts and constructs a plan to overcome them and succeed despite understanding that it may be difficult to achieve. And all this despite the naysayers, who always exist and are in another category.

Our industry is at a crossroads. In the future, our industry's landscape and the economic global landscape will change. It is inevitable. This new reality will require a few good and committed leaders to set the example. Financial institutions will have to be clear about how they are distinctive and make these differences more obvious with every human interaction. We also have to face the fact that our basic model is under pressure and that the small incremental improvement approach will no longer be enough. A more substantial change to how we operate, market and package our products and services will have to take place. Because almost 70 percent of our operating costs consist of people and systems, we need great people and best-in-class systems. A technology partner with a flexible, low cost architecture can provide the flexibility required to navigate the next decade or two, as will having the right people to make the systems work.

It seems clear that this will take extraordinary effort by a few extraordinary people. They will be the leaders who inspire an industry and excite the senses about what is possible. In the end, they will lead the industry back to the center of economic vitality and reposition financial institutions back to their natural role of trusted financial intermediaries.

I hope you will find it within you to commit to becoming one of these people. It only takes a small number of extraordinary individuals to show the way.

LaVergne, TN USA
06 April 2010
178265LV00002B/2/P